ABOUT THE AUTHOR

Denise Caruso is the co-founder and executive director of The Hybrid Vigor Institute, a not-for-profit research and consulting practice based in San Francisco. Her most recent work has focused on new, collaborative methods for addressing the risks of innovations in science and technology, as well as new approaches to the risks of global infectious disease. Also a veteran technology journalist and analyst, she began covering the personal computer era in the early 1980s for a variety of trade and national publications. For the five years prior to founding Hybrid Vigor in 2000, Caruso wrote the Digital Commerce column for *The New York Times.*

INTERVENTION

INTERVENTION

Confronting the Real Risks of Genetic Engineering and Life on a Biotech Planet

DENISE CARUSO

The Hybrid Vigor Institute
San Francisco, California

Published by Hybrid Vigor Press
The Hybrid Vigor Institute
San Francisco, California 94107

Caruso, Denise
Intervention

Includes bibliographic references and index
ISBN 978-0-6151-3553-3

Intervention is available via
http://hybridvigor.org
http://hybridvigor.net

For my mother and father, who taught me never to be afraid of a hard problem.

Intervention: a planned interaction, with the aim of making a full assessment, overcoming denial and interrupting a harmful behavior. The preferred technique is to present facts regarding the behavior in a believable and understandable manner.

CONTENTS

AUTHOR'S NOTE

REGARDING AMATEURS, STEREOTYPES AND SOURCE MATERIAL

To develop the arguments at the heart of this book, I sampled the work of experts and scholars from many different disciplines. As a result, I expect some of them will feel that having a generalist like me traipsing through their fields does their work a disservice.

A well-known historian had similar concerns; in his preface to *The Columbian Exchange*, about the biological and cultural consequences of Columbus' discovery of America, Alfred Crosby noted that scientists and scholars in the several fields he consulted would likely see him as "an amateur." After agreeing with them in part, he responded that "there is great need for... pulling together the discoveries of the specialists to learn what we know, in general, about life on this planet."[1]

I don't presume to compare my work to Crosby's, but this "pulling together" of ideas and discoveries in service of the bigger picture is the purpose of *Intervention* as well. I've been as scrupulous as possible about representing others' work fairly, but I take full responsibility for any misinterpretations.

I also want to apologize in advance for what will appear to be blatant stereotyping of science and scientists, regulators and people who work in the biotech industry. I have deep respect for people who have dedicated their lives to expanding the bounds of human knowledge and helping humankind, whether in private industry, academia or government. I mean no offense to those who don't fit what I know are overly broad descriptions. It would have been impossibly unwieldy to include a disclaimer every time I used them.

Finally, readers will notice an abundance of numbered references throughout the text. All are linked to source material, listed by chapter in the back of the book. There is no need to flip back and forth while reading unless compelled to know a reference at that very moment. Any comment that is important to the text itself will be footnoted on the same page.

INTRODUCTION
(GENE)SIS

The seed that yielded this book was planted the day I received an e-mail out of the blue from Roger Brent, a biologist who runs the Molecular Sciences Institute in Berkeley, Calif. MSI is an independent laboratory, founded by the Nobel Laureate Sydney Brenner. Its research is focused on what Brent calls "predictive" biology, discovering how cells work way down deep, wanting to know them with the same intimacy that a good car mechanic knows every cog and piston of an engine.

To give you an idea how far away biologists are from achieving this level of prediction, and how complex the problem really is, in 2002 Brent's lab was awarded more than $15 million by the U.S. National Human Genome Research Institute to spend five years studying one communication pathway in one cell — that cell being baker's yeast, an organism that, I feel compelled to point out, only *has* one cell.

Brent and I became friends, and it wasn't long before he was razzing me, albeit kindly, about what he assumed was some new-age, crunchy-granola California attitude on my part about transgenic food, which I continue to avoid whenever I have the choice. (The more common term is "genetically modified" — a.k.a. GM — or "genetically engineered." But because any domesticated plant or animal has by definition been crossbred for desirable traits, and thus technically has been "genetically modified," the proper term is "transgenic.") Crops that have been engineered to include genes from *Bacillus thuringiensis*, or *Bt*, a soil bacterium, are toxic to certain kinds of insect pests, he assured me, but not to humans.[1]

"You know, you could eat 10 kilos of *Bt* potatoes and not get sick," Brent said.

"You *don't* know that," I countered. "I may not get sick today, but you don't know what happens next. You don't know everything that happens to those engineered genes once I eat them. You don't know if they're having effects on me other than making me sick. You don't know what they might be doing to other animals that eat them, or to other plants or the environment. Whether I get sick is almost beside the point, in terms of what else could be happening that *you guys don't know anything about.*"

I was surprised that he concurred. Then he asked, "So what do you suggest we do about it? How do we not stop science and protect people at the same time?"

I thought for a moment. "Well," I said, "I think we have to redefine risk."

In that moment, I realized that it was our individual perceptions and definitions of *risk*, not our knowledge (or lack of it) about genetic engineering or biology, that had created the rift between my worldview and Brent's. I question the safety of genetic engineering based on my perception of how much we don't know about it. Brent is confident that it's safe, based on his perception of how much he and his tribe of molecular biologists *do* know. And because he's the scientist and I'm not, he wins the point by default.

This is precisely the skewed dynamic that governs how the risks of innovations are assessed and otherwise regulated in our sophisticated technological society. We've been taught and trained to assume that science is truth and that there's a scientific answer to all our concerns about risk. But in reality, there may not be just one answer, and science may not be able to provide the answer that's most relevant to our concerns.

≡

Risk-taking has been a conscious human activity for nearly as long as we have records. In fact, risky behavior in the form of gambling may trump prostitution for the title of Oldest Human Vice. Cave paintings show our ancestors playing a game of chance, with dice made from the ankle bones of deer and sheep, long before the bestowing of sexual favors became a commodity. But our gambling ancestors had no conception of the literally calculated risks of today. It took thousands of years from the time that first ankle bone was thrown for *Homo sapiens* to develop the tools that allowed us to calculate the probability that a given event would happen again, based on how often it had happened in the past.

This "taming of chance," as one philosopher called the mathematics of probability, dramatically changed how humans saw their role in the natural world. For the first time, they could exercise some control over their destiny. With a systematic method to forecast past hazards that would occur again, taking risks became less personal: a pending danger was less about the wrath of God and more about the luck of the draw. This newfound skill was so transformative that virtually all the technological progress of the past century — affecting everything from human health and longevity to environmental regulations, financial markets, industrial development and growth — can be attributed in some way to the ability to mathematically assess and forecast certain kinds of risk.

As a matter of course, technology can bring us into daily contact with three different kinds of risk.[2] One is the risk of impending danger: a random or unpredictable hazard with catastrophic potential. We often connect this type of risk to large-scale technologies, like nuclear power. The second category might be called the risk of long, slow poisons: invisible threats we can't see with the naked eye and that we can avoid only if we're told to — by experts, like scientists, who are supposed to know about them, or by those who are entrusted to protect us from them, like government regulators. Here reside risks like food additives, and chemicals in pesticides, drinking water and various plastics. Third are the risks that present themselves in the form of a cost-benefit ratio. These risks are primarily thought of as financial, but even when they aren't, most everyone is subjected to the balancing act they require: the public, if they have the choice, deciding whether the cost or risk of a new technology or health intervention is worth the promised benefit; regulators deciding whether a risk is sufficiently threatening to justify the cost of additional bureaucracy; and industries deciding whether they can make enough profit to justify the research, development and marketing costs of selling a new technology to the public.

The method for determining the reality of these threats and justifications, as accurately and effectively as possible, is called a risk analysis. Its purpose is to help us make better and more rational decisions in the face of uncertainty. In the years before the publication in 1962 of *Silent Spring*, Rachel Carson's landmark book detailing the unintended effects of DDT on health and the environment, the mathematics of probability was the only real analytical tool available for determining risk. Back then, the field was dominated by engineers, economists and epidemiologists, who calculated risk based on historical data and knowledge of their own subject areas. But the magnitude of the DDT disaster and other disasters that followed — the Chernobyl and Three Mile Island nuclear accidents, the *Exxon Valdez* oil spill, the *Challenger* shuttle explosion, the chemical spill at Bhopal — challenged the value of these mathematical analyses and instigated what has become an ongoing debate about how to assess the unique risks that science and technology engender.

The central issue, as seen by concerned scientists, policy makers, citizens and other stakeholders, is scientific uncertainty: how to go about properly assessing the risks of complex technical systems, given how little is known about them or how they interact with each other. Relying solely on scientific "evidence" or mathematical assessments of risk, the critics say, has led regulators of innovations to make safety determinations by asking only the questions that scientists can answer — and ignoring the larger, more complex questions that they cannot.

That explains, at least in part, why most of what we hear or read in the popular press about genetic engineering and biotechnology controversies is framed as all or nothing — all promise or all peril — and why people stake out such evangelical positions at the extremes of pro and con. They aren't just at cross-purposes. They aren't even on the same topic! Those on the side of all promise see only the "evidence" of the benefits of transgenic food, for example, and if you don't agree, then you don't care if the starving little children die. Those who feel imperiled see only the uncertainty, believing that uncertainty itself is evidence that genetic manipulations are certain to go awry — and if you disagree, you don't care if the planet winks out in some cascading biological death spiral. And we ordinary people are stuck in the middle, not knowing whose brand of fundamentalism we should believe.

≡

I have no axe to grind with biotechnology. I'm not opposed in principle to biotech or to any kind of technology, for that matter. I spent nearly 20 years of my career deep in the heart of Silicon Valley, as a journalist and analyst and commentator on all manner of things that used to be called "high technology." I am comfortably intrepid when it comes to such matters, and I appreciate and enjoy the fruits of progress.

But I am also a skeptic red in tooth and claw. And I've been around the block enough to know that all technologies are double-edged. I watch my back when anyone makes extravagant claims about safety and benefit but doesn't show me how this sublime state of confidence was reached. I am wary when people won't engage in a meaningful discussion about risks and drawbacks but instead dismiss my concerns as ignorant or ill-informed or "not science-based." I am equally dubious when someone tells me that something is 100 percent evil and must be stopped immediately at any cost. I have also learned, as has anyone who reads the business news these days, to follow the money if I want to understand how and why the world works.

With these signposts in mind, I began sifting through the claims of the cheerleaders and the declamations of the activists. I ferreted out the less vocal stakeholders and the scientists whose work and perspectives defied pro and con polarization. I read hundreds of documents on genetic engineering and transgenic organisms and genomics. I combed through reams of social science literature on risk and probability and science and public policy. After all that, I came to the conclusion that I don't know enough to decide what's risky about the products and processes of genetic engineering.

And neither does anyone else.

That's all I know for sure. *Nobody knows.* No one person or group knows or understands enough about the complexity of living things or their intimate interactions or what affects them to declare that biotechnology and genetic engineering are risk-free. In fact, the only thing we all share — scientists, citizens, regulators — is the profound uncertainty of this moment in history. Recall the words of Craig Venter, the former president of Celera Genomics, just after he led one of two competing teams to finish sequencing the human genome in 2000:

> With this technology, we are literally coming out of the dark ages of biology. As a civilization, we know far less than one percent of what will be known about biology, human physiology, and medicine. My view of biology is 'We don't know shit.'[3]

At the very moment Venter was making this lyric statement, farmers were preparing to plant more than 82 million acres of transgenic crops in the United States. A survey of that year's crops by the U.S. Department of Agriculture (USDA) showed that 68 percent of all soybeans planted in the U.S. were transgenic, as were 69 percent of all cotton crops and 26 percent of corn acreage.[4] Those percentages are increasing at a rapid clip in countries around the world.

Meanwhile, the next generations of transgenics are being developed and prepared for market. Far more radical than today's transgenic farm crops, they include pigs engineered to provide transplant organs for humans, ultra-fast-growing fish, microbes designed to "eat" pollution, hypoallergenic cats and viruses re-engineered to burst cancer cells, to name but a paltry few.

If this is the result of knowing "far less than one percent of what will be known" about biology, I fear for the future. The living planet has already become a giant genetics experiment. No one knows how many millions of transgenic organisms are out there in the wild now. Uncontrolled and unmonitored, they are simply doing what living things do: reproducing and evolving and sharing their genetic material with all the other complex forms of wild and engineered life that inhabit the earth. And despite all official reassurances of safety, no one actually knows what the wild and the engineered are eventually going to make of each other — or maybe, eventually, of us.

CHAPTER 1.
WHAT IF THE EXPERTS ARE WRONG?

"The question is, how do you prepare to be wrong? If you know you can't walk away from the consequences of what you do, how do you not screw it up?" said Todd La Porte, sitting across from me in the dappled light of the faculty dining room at the University of California, Berkeley. La Porte, a former Marine, is also a veteran political scientist who is internationally known for his thoughtful study of "long-term stewardship" of man-made hazards; that is, how a society prepares to take care of the messes it has made that it can't get rid of, generations into the future.

La Porte has spent many years studying how nuclear engineers and scientists go about the business of containing radioactive waste, which to date is the most persevering toxic substance known to (and created by) man. I had contacted him when I first started my research into risk and genetic engineering. It occurred to me that if something bad happened as a result of our self-assured release of transgenic organisms throughout the world, we might eventually need to have a more intimate understanding of his work. For starters, as La Porte noted, "nuclear waste doesn't reproduce." A population of living, multiplying transgenic organisms gone awry could end up being significantly more difficult to contain than radioactive sludge.

Should such a thing happen, it would create stewardship challenges for generations into the future that are already far beyond our present scientific knowledge or capabilities.

And while the thought of being wrong about having stocked the entire planet with self-replicating hazards was sobering enough, La Porte posed yet another, equally troubling question about the topic of my inquiry: "How are you going to get the scientists to listen to you?"

After decades of study, La Porte himself had no answer. "My experience with technical people, with scientists, is that they're utopians and they see *us* as the problem," he said. "This was a tragedy in the nuclear industry."

Nuclear scientists, said La Porte, entered their profession believing they were doing something good for the world by developing what was then called "atomic energy." Many of us remember this era, when nuclear energy was pervasively (and now infamously) touted by the nuclear industry as 100 percent safe, clean and "too cheap to meter." The scientific basis for those claims was accurate as far as it went, but clearly it didn't go far enough. When the industry's claims of safety literally blew up — with operator and engineering errors triggering the meltdowns at Three Mile Island in 1979 and Chernobyl in 1986 — the public rejected the technology as too risky for the benefits promised by its government and industry champions.

A new focus on our global dependence on fossil fuels has some people trying to salvage nuclear energy's reputation. But nuclear power plants are still considered too vulnerable to human error to be reliable, and the issue of how to safely store and safeguard radioactive waste remains a weighty and so far intractable problem, for both public health and global security.

"To think that other people might suffer as a result of their actions is not part of the expert's world, or it gets pushed away in the drive to deploy the technology," said La Porte. "But what are the consequences if it turns out that all the things they believed in are wrong? That's really hard. And most technical people can't talk about this. What they do is theology to them, not science.

"This attitude used to be less of a problem because we couldn't destroy the earth," La Porte continued. "But now we can. The consequences of error are likely to be greater than they were. The power of technology is much greater. The capacity for untoward error is much greater. And the effects of technology at scale have not been tested in the area of biology."

≡

This untested technology is, of course, biotechnology. Using a laboratory technique known as "recombinant DNA," scientists now can splice together the genetic material from deep within the cells of two or more organisms of different species. As a result, they can "engineer" living hybrids with new traits that would have been impossible to create using traditional breeding techniques.

Genetic engineering commenced what was heralded as a new era, both of scientific discovery and commercial potential. The technique itself was quickly patented, and the first biotech company, Genentech, Inc., was launched in 1976. A torrent of research and experimentation followed, and a new generation of genetic engineers immediately began to add, remove or otherwise modify the DNA of all kinds of living things. The term "genome" had long since been understood to describe all the genes in an organism. But this newfound ability to directly manipulate individual DNA sequences to change the way that organisms behave provided new impetus to discover and map as many genes and their functions as possible. In 1977, for the first time, the entire complement of genetic material in a biological entity — a virus that kills bacteria, called a bacteriophage — was mapped and published.[1]

Many more genomes were mapped and published in subsequent decades. But the climax of these efforts was the dramatic completion of a working draft of the human genome map in June 2000.[2] For many people, this historic achievement — combined with the power of recombinant DNA to re-engineer the structures and behaviors not just of microbes, plants, and other animals, but of humans as well — inspired researchers to dream big about how humankind could use this knowledge.

And like La Porte's nuclear utopians, dream big they have done. Over the past several decades we've heard an ongoing stream of promises about how genetic engineering and its products are on their way to eliminating infectious disease, ending world hunger and even repairing the tremendous damage we've wreaked upon the biosphere.

But what we know from history is that every promise based on discovery or invention, no matter how positive, comes factory-equipped with its own unintended dark-side consequences. For all the utopian results that genetic engineers have imagined for us, the ability to "rewire" the genetic material of living organisms could just as plausibly yield an equal and opposite nightmare. It is not especially difficult to come up with scenarios whereby mucking around in the genes of living organisms leads to serious biological, social and/or economic disruption. Yet neither knowledge of history nor dark-side scenarios has tempered the zeal or the speed with which the products of genetic engineering are being dispatched into the global marketplace.

Are the experts who build these products thinking critically about these dark possibilities? What set of facts, based on what specific scientific knowledge, have they provided to government regulators who decide whether the products of genetic engineering are safe? Do either the scientists or the regulators know enough about what they're doing with this largely unexplored science to speed biotech products to market as quickly as they are today?

As sensible and rational as these questions sound to most of us, it turns out that this is a most unwelcome line of inquiry from the expert perspective. Ask the people whose livelihoods are intertwined with science and technology about the risks of what they do or sell — particularly if they are at or near the levers of power in academia, industry or government regulation — and you'll generally get a scorching look of suspicion and almost always at least one (and sometimes all) of the following three reactions:

"People are ignorant. This technology is absolutely safe."

"The public is scientifically illiterate. There's no point involving them in the conversation; they just get scared and stop us from doing our work."

"The problem is that people just don't understand risk."

On the surface, there's no denying public ignorance. Making that case is like shooting fish in a barrel. Not many people even know how their televisions work, let alone how scientists can "engineer" the DNA that resides deep within the cells of all living things. But this is a terribly elitist argument. As the Canadian philosopher John Ralston Saul wrote, "When faced by questioning from non-experts, the scientist invariably retreats behind veils of complication and specialization, [making] it impossible for the citizen to know and to understand, and therefore to act, except in ignorance."[3] What's more, the claim that ordinary people are incapable of understanding the risks of scientific and technological interventions has been proven to be patently untrue, time and again, by risk researchers.

This tacit refusal to directly address the public's concerns about risk obscures many larger and more profound truths about the process of scientific inquiry and discovery, truths that are rarely acknowledged in the context of how this process affects expert assessments of risk. Using the false pretense of public ignorance as a shield turns the public's legitimate — and generally quite relevant — questions into a dangerous game of "us versus them," when far more complex factors than public ignorance are at play.

≡

To begin with, those who discover, invent or work with new technologies are often spectacularly nearsighted about the risks those technologies create. To deny this is to ignore at least a century of the history of biology and technological advancement.

The tragedy of the drug DES, for example, continues to reverberate through generations. As many as 10 million pregnant women in the U.S. alone took diethylstilbestrol, a synthetic estrogen, between 1940 and 1971 (despite several studies that proved its ineffectiveness) hoping that it would prevent miscarriages.[4] But in 1971, researchers discovered the link between daughters of DES mothers and what was until then a rare cancer: clear cell adenocarcinoma.[5]

Animal studies a decade earlier had signaled possible links between early estrogen exposure and later cancers in offspring, yet these findings had been dismissed and considered irrelevant to human health by doctors and drug makers, as well as by the U.S. Food and Drug Administration (FDA), which had approved DES. But even after human studies made the linkage irrefutable, researchers had to fight to get colleagues in the scientific and medical communities to believe the proof. Remarkably, the skepticism continues, even as many more problems have surfaced in the subsequent decades, some of which also affect sons of DES mothers. Research now shows that even the children of DES children are at high risk for cancer and other DES-related health problems.[6]

Other, similar long-term disasters may be looming as researchers are discovering serious health implications for people exposed to possibly the most ubiquitous man-made substance in our lives: plastics.

Studies were released in 2003 on three specific plastics* that are found in virtually everything around us: food containers and cooking utensils, clothing, cars, furniture, medical and dental appliances, paint, and even bubble gum. Population samples show that as many as *92 percent* of Americans may have traces of one of them, PFOA, in their blood. The problems that all three of the chemicals are suspected to cause range from kidney, reproductive and brain development problems to thyroid cancer.

---

*Perfluorooctanoic acid (PFOA), bisphenol A (BPA), and polybrominated diphenyl ether (PBDE).

The millions of children exposed to DES pales in comparison to the multitudes worldwide who have been exposed to any one, let alone all three, of these plastics.

As has become standard practice, industry groups representing the manufacturers of these chemicals continue to insist that they are safe, and hold up as "proof" the fact that regulatory agencies have not yet taken action against them.[7,8,9,10] (One exception is PBDE; in 2004, the U.S. Environmental Protection Agency (EPA) finally negotiated a phase-out of this chemical as a result of pressure by consumer groups.[11])

Another similar health crisis is already well under way as a result of our overuse of man-made antibiotics — namely, the steep increase in antibiotic resistance that many dangerous pathogens have developed as a result.

Antibiotics were once considered miracle drugs that, for the first time in history, greatly reduced the probability that people would die from common bacterial infections. But once these new drugs became cheap and readily available, doctors prescribed them for virtually every ailment, often thoughtlessly or incorrectly. As a result, bacteria became immune to the drugs that once killed them.

Resistance to antibiotics has become pervasive among pathogens that infect people and animals all around the world. In hospitals in particular, patients often contract "superbugs," like *Staphylococcus aureus* or pneumonia, which now are virtually unkillable. Staph infections, for example, are already resistant to common antibiotics like penicillin, methicillin, tetracycline and erythromycin. As a result, these low-cost treatments have become practically useless for common infections. This leads to more frequent use of newer and more expensive compounds, which in turn leads inexorably to the rise of resistance to the new drugs as well. A never-ending, ever-spiraling race to discover new and different antibiotics has ensued, just to prevent losing further ground in the battle against infection.

The situation is worsened by the fact that the genetic material responsible for conferring antibiotic resistance can move with relative ease between different species of bacteria. This is evolutionary selection in action: the transfer of resistance makes it possible for pathogens never exposed to an antibiotic to acquire resistance from those that have been and thus survive. (Antibiotic-resistant genes play an important role in genetic engineering as well, as you'll see.)

Another great concern is that in the United States, antibiotics are still routinely included in the diets of healthy livestock, for no reason other than to make the animals grow faster. But now the bacteria the animals harbor have become widely resistant to antibiotics, too. It has been well documented by the U.S. Centers for Disease Control and Prevention (CDC) and by the U.S. FDA that since the time these farm animals were first fed medically unnecessary doses of antibiotics, the meat supply has become highly contaminated with bacteria. What's more, foodborne illness has become a much more serious problem — especially illnesses caused by *Salmonella*, *Campylobacter* and *E. coli*, pathogens that are resistant to nearly all antibiotics. In addition to the issue of foodborne illness from contamination, the resistant bacteria get passed along to humans who eat resistant animals, like chickens and cows, or their products, like eggs and milk.[12]

As a result of this growing problem, many countries have long since banned the use of antibiotics for growth promotion or disease prevention. In the U.S., however, it took until March 2004 before the FDA disallowed just one single type of antibiotic — enrofloxicin — that was widely used in poultry.[13] Enrofloxicin in animals metabolizes into ciprofloxicin, a.k.a. Cipro, the drug that made headlines in 2001 as the treatment of choice for humans who inhale anthrax spores.[14]

≡

Scientific shortsightedness does not apply only to products, but to discoveries as well. It can be hard to measure how much scientific progress is held back by a research community too mired in its prejudices to accept truly revolutionary discoveries. The history of science is full of such examples, but a relevant one for the 21st century is the story of Stanley Prusiner, a neurologist at the University of California, San Francisco. Prusiner lost much of his funding, his academic tenure (temporarily), and, for many years, credibility in his field — all for research that would later bring him a Nobel Prize in Medicine. In 1982, he discovered a strange misfolded protein that he called a "prion" (he derived the word from protein and infectious[15]) and that apparently could transmit disease.

In fact, it is now widely known that prions *can* transmit disease. They are the infectious agent that causes the brain-wasting disease in animals known as transmissible spongiform encephalopathy (TSE). In its variant forms, it's known as bovine spongiform encephalopathy (BSE) or mad-cow disease in cattle, scrapie in sheep and "variant Creutzfeld-Jacob Disease" (vCJD) in humans. TSE was also recently discovered in goats and deer, and it affects other animals as well, including squirrels, elk and mink. In fact, an ongoing TSE epidemic in Colorado and Wyoming is exposing both cattle and hunters who share the same terrain to TSE. No human or bovine infections have been documented, but our lack of understanding of the disease and its long incubation period — it can take from three to 50 years for symptoms to appear — continues to be worrisome to many, despite reassurances from the USDA.

The research community had believed that these brain-wasting diseases, which Prusiner had traced to prions, were caused by viruses. A virus is a parasite with no cell of its own, so it has to "hijack" the DNA in the cells of another organism in order to reproduce and become infectious. A virus can do this because it contains the machinery of reproduction — i.e., DNA or the RNA molecules that help decode the information carried by DNA. But prions do not contain DNA or RNA. In fact, prions are the only known infectious agents that *don't* contain DNA or RNA. Bacteria contain DNA. So do fungi, parasites and protozoa. So when Prusiner isolated the infective protein particle, scientists simply refused to believe that it contained no genetic material. In fact, more than 20 years after his discovery, some definitions still call prions infective particles "which (almost certainly) do not have a nucleic acid genome."[16]

Despite Prusiner's prior achievements, many scientists in the research community also discounted his more recent claims that prions reside not only in the spinal cord but also in the muscle tissue of animals that we eat. Yet in 2006, prions had been, in fact, discovered in many other parts of animals, including the muscle tissue of North American deer and elk.[17] And these scientists are presently rejecting his ideas about the relationship of prion diseases to other disorders, such as Alzheimer's and Parkinson's diseases. Time will tell who prevails.

Scientists don't hurt only themselves with this kind of behavior. They hurt us, too. By refusing, for whatever reasons, to look beyond the narrow boundaries of their own expertise, they often have overlooked the cause of problems as well as potential cures or solutions. Prusiner himself best sums up what scientists risk by indulging in these dogmatic attitudes: "While it is quite reasonable for scientists to be skeptical of new ideas that do not fit within the accepted realm of scientific knowledge," he wrote with great understatement in his Nobel autobiography, "the best science often emerges from situations where results carefully obtained do not fit within the accepted paradigms."[18]

≡

What's even more distressing is how frequently scientists reject these "results carefully obtained" when they actually do fit within the bounds of paradigms they understand. This type of scientific myopia may be closest in spirit to the issues in question around genetic engineering. It is also where we find what may be the most persistently damaging effect of shortsightedness: the destructive and exponential growth of invasive species.

An invasive species is any plant, animal, microbe or virus — including any of the biological components that can propagate them, like seeds, eggs, spores or infectious bits — that's not native to a given region or ecosystem and that out-competes native species for space and resources.

While humans are almost always responsible for such invasions, not all invasive species are purposely introduced. They've been known to stow away in ballast water of ships, in crevices of airplanes, or attached to the clothing or shoes of unsuspecting travelers. One of the most rampant accidental invaders was never meant to live outside a fish tank. This human-bred "aquarium strain" of green algae called *Caulerpa taxifolia* has already literally choked the life out of tens of thousands of acres of seafloor around the world.[19]

Everything that was good about this plant in a fish tank turned out to be an ecological disaster in the wild. It was bred to tolerate cold water and it grows fast (up to a full inch per day). Because even a small broken-off fragment can form a whole new plant, it hitches a ride on boat anchors and fishing gear and quite easily starts up new colonies at great distances from the original source. What's worse, outside of the native strain's tropical home, no fish will eat it because it produces toxins that taste bad.

There's a lesson here, both about the banality of evil and about the consequences of technology at scale, which should be carried forward into any conversation about genetic engineering. Who would have imagined, before *Caulerpa*, that such far-reaching damage could be caused by an activity as everyday as dumping the contents of an aquarium down a drainpipe?[20]

People make mistakes and accidents happen, of course; unintended infestations by alien plants, animals and microbes are one of the risks of global mobility. But even more distressing are the invasive species that were purposely introduced. Time and again, various government agencies and people with the best of intentions, having had quite enough of one pest or another, have imported various critters to combat them. These immigrants, deliberately introduced, often become pests far worse than those they were brought in to eradicate.

One notorious example is the Hawaiian cane toad, *Bufo marinus*, brought into Australia in 1935 to rid its sugarcane plantations of cane beetles. The brains behind this idea was the Australian Bureau of Sugar Experimental Stations, which apparently didn't ask *Bufo* for references before hiring. The fact is that this toad has an immense appetite for everything *but* the cane beetle. It is big and aggressive, and its skin is poisonous to any natural predator — except one lone snake species, which is destined never to hunger again. Worse yet, the tadpoles of cane toads mature earlier than other tadpoles in Australia, so in addition to being nasty-tasting to potential predators, the hungry babies also eat up everyone else's food.

With these unnatural advantages, it didn't take long for the toads to spread along the north coast toward the center of the continent, eating all the native amphibian and invertebrate species in their path — except, as noted, the cane beetle, which flies over its head, and cane grubs. While the grubs are at least within reach, the toads apparently cannot be bothered with them, since they live below the soil and require at least a token amount of effort — effort that's quite unnecessary given the toads' luxurious circumstances.

Similarly, the introduction of the European rabbit to Australia as a game animal proved to be a mistake of magnificent proportions, and the proposed solutions are proving even more frightening than the original invasion. By most historical accounts, 24 wild rabbits arrived in Australia from England on Christmas Day in 1859; 10 years later, some 2 million per year were being shot with no noticeable effect on the population. The original colony may have produced more than half a *billion* rabbits on the continent, destroying vast tracts of vegetation and contributing to the extinction of many native marsupial species, like the bandicoot.

If we were looking for evidence of scientific hubris, we need look no further than two subsequent steps in the rabbit saga. Australian scientists decided they would try importing diseases into the rabbit population to kill them. Both of the diseases they were considering as imports were also not native to Australia and were supposed to be spread only by rabbit-to-rabbit contact. Instead, as the scientists found out wholly by accident, what spread the diseases (myxoma in the 1950s and rabbit hemorrhagic disease virus in 1995[21]) were biting and stinging insects such as mosquitoes. For some unfathomable reason this was unforeseen by researchers, despite the fact that mosquitoes and other similar biting and stinging insects are among the best known disease vectors in the world. Luckily, the two diseases weren't like West Nile; that is, transmissible to humans or other animal populations by mosquito. Otherwise, the scientists might have had an epidemic on their hands.

A cautionary tale written by two scientists notes that the two attempts to introduce diseases to try to control the rabbits are distressingly similar:

> ...In each case, because of errors in our assumptions, the organism successfully escaped rather than being [purposely] released ...The pattern of invasion did not match that predicted from available information, as the proposed mode of infection was not that used. The speed of the spread was also much faster than expected and was unstoppable.
>
> The unexpected behaviour of all three species places a serious question mark over our ability to predict the behaviour of invading organisms placed in ecologically different environments, and thus to protect a naive environment from an invasion. The lessons that should have been learned from the escape of myxoma were not accepted, and RHDV escaped in a virtually identical manner. Have we learned our lessons yet, or can we expect similar escapes in the future?[22]

Based on recent history, we can hazard an answer to both those questions: No, we haven't; and yes, we can.

At various times Australian scientists have tried to import viruses from other countries, including Venezuela, as biological controls for cane toads as well — which, considering that the toads were intended to be a biological control themselves, is something akin to fighting fire with gasoline. A research organization established nearly a century ago to benefit Australian industry claims to be working with "cutting edge genetic technology" to find a biological control method to "stop the hop" of cane toads across the continent.[23] These types of efforts continue despite the fact that in 2001, researchers in Canberra, working to create a genetically engineered sterility vaccine to control a national infestation of mice, instead accidentally created a "mousepox" virus so powerful that it killed even the mice that had been inoculated against it.[24] (A U.S. team of researchers immediately replicated it, in the name of biological defense.[25])

While public ignorance is most often cited as the reason that risk is so misunderstood, these are examples of *scientific* ignorance. In each of these situations, the proposed intervention was subjected to some degree of regulatory and/or scientific scrutiny. Each received some degree of pushback from concerned scientists or the public that questioned assumptions of safety — often in the form of data that refuted their proposed actions — and those involved in making the decision had an opportunity to revise their "beliefs" (I use the term advisedly). Instead, they ignored the pushback and declared the interventions to be safe, or safe within what were believed to be easily defensible and understood boundaries.

≡

Another factor in the public's relationship to risk, which one could probably call *unintended* public ignorance, affects us more often than we can know. Unintended ignorance results when regulatory agencies or industries willfully downplay or deny the risks that are already known to them, in the interest of protecting financial or some other kind of gain.[26] In the early 21st century, this game of "hide the risk" has already reached epidemic proportions in the United States at least.

Public-interest groups in the U.S. have railed for decades about the dangers of the revolving door between government and industry, whereby people with a financial interest in a given industry — industries that generally provide largesse in various forms to those in power — are asked to serve as regulators of that industry. The practice has become increasingly common and bold, and as a result, American citizens are witnessing an ongoing rollback of hard-fought federal safeguards in agencies that regulate food safety, the quality of drinking water, worker health and safety, civil rights, toxic pollution, health care and other common public resources.

The most blatant recent examples in the U.S. have involved the government censorship of EPA reports that connected auto emissions and other human activities with global warming; EPA administrators selectively editing a risk analysis that the agency commissioned on mercury emissions; the sabotaging of a World Health Organization initiative on obesity because the sugar and packaged food industries felt "attacked" and opposed its suggestions; and the stacking of a CDC committee with industry-friendly experts to re-examine federal standards for lead in school drinking water.[27,28,29,30]

With just these few various historical examples in mind, it's not surprising that people don't trust the information supplied by their governments or the scientists who advise them about the risks and benefits of genetic engineering. If we want to get the straight story, we need to look closely at two key areas. First, we must ask whether we are getting the whole story about how much scientists truly understand the biological processes that they are altering via genetic engineering. Second, where risk itself is concerned, we need to ask if we are getting the quality of analysis we deserve from our government regulators about those genetic alterations and the biotech industry that sells the products that result.

CHAPTER 2.
OF MICE, MEN AND UNCERTAINTY

Once James Watson and Francis Crick had famously solved the structure of the DNA molecule in 1953, the graceful spiral of the double helix became synonymous in almost every context with what today we call "genes." The belief that DNA was the chemical stuff of heredity, that it explained why and how the traits of living things can be consistently similar from generation to generation, seemed so conclusive that DNA was declared "the blueprint of life."

While it's inarguable that DNA can explain some of the most distinctive aspects of heredity, it's certainly no blueprint. If it were, it could explain what continues to be one of the greatest puzzles in biology: how can organisms that share all or much of the same genetic makeup look and behave so *differently*? This question is puzzling enough when the organisms are the same species. Why and how do identical twins, for example, or seeds harvested from the same plant, grow up to be so different from each other?

The comparisons across species are even more confounding. A mouse has about 25,000 genes, roughly the same number as a human. Our cells and organs operate so similarly that millions of mice have been used in laboratories to study how genes work, as well as to test the effects of drugs and other types of therapies. So how is it that humans can be so genetically similar to a tiny, furry, four-legged rodent — with a tail and the ability to birth a litter of pups every four or five weeks? Clearly the reality of heredity demands a far different explanation than DNA alone can provide.

Indeed, "no one who has studied the genome ... could be anything but humbled when the human and mouse genomes were aligned and compared," said Shirley Tilghman, the president of Princeton University. The renowned molecular biologist was speaking at a scientific meeting in 2003, celebrating the 50th anniversary of Watson and Crick's discovery. "Clearly what is immediately apparent when you look at any part of those two genomes ... is that evolution has indeed been hard at work conserving far more of the genome than could be explained by genes."[1]

Tilghman's observation may be humbling to scientists, but to people who are concerned about the unchecked advance of commercial transgenics, it is deeply troubling. Over and over we have been told by scientists of Tilghman's caliber — many of them winners of Nobel Prizes — that there is no scientific evidence that genetic engineering is risky, that the concerns raised over the decades are not "science based."[2] But what does this lack of evidence mean in the context of what they don't know? If Tilghman and the rest of the best molecular biologists in the world can't explain the genetic similarities between mice and men, how can they claim to know enough to tinker with the DNA of living organisms — and, at the same time, declare there is no scientific evidence for risk? How can they admit to so much uncertainty *without* acknowledging risk?

≡

The factors contributing to this state of affairs are layered and complex, rife with political, economic and cultural conflicts. The ancestor of them all is the nature of scientific evidence and expertise, and for genetics the source is fairly easy to trace. It goes straight back to Gregor Mendel, the father of modern genetics.

Mendel was a 19th-century Austrian monk who, in his spare time, bred and studied pea plants in the monastery's gardens. He was able to control the parentage of each generation of plants using that proto-genetic engineering technology called "breeding," whereby he carefully pollinated the flowers himself, rather than allowing nature to take its usual course, à la the birds and the bees. This careful control allowed Mendel to produce unequivocal evidence that certain distinct traits in the pea plants, passed to subsequent generations, were determined by some specific factor — some "unit" of heredity. Although he didn't know exactly what this factor was (the term "gene" wasn't coined for several more decades, and the DNA molecule was not shown to carry hereditary information until nearly a century later),[3] Mendel could still use what he'd learned by observing how these traits were reproduced. As a result, he was able to selectively and predictably breed smooth and lumpy peas, tall and short plants, white and purple flowers, and so on.

For the purpose of his investigations, Mendel viewed the plants as if they were simply a collection of what he called characters, or traits. But he didn't study *all* of their characters or traits. At the beginning of his famous 1866 article on genetics, Mendel stated plainly that, as a condition of his experiments, he didn't include such traits as the form of the plant or the size of its leaves.

Why? Because they were simply too difficult to study.

"Some of the traits listed do not permit a definite and sharp separation, since the difference rests on a 'more or less' which is often difficult to define," Mendel wrote. "Such traits were not usable for individual experiments; these had to be limited to characteristics which stand out clearly and decisively in the plants."[4,5]

What Mendel didn't realize was that by limiting his observations to distinctive traits, he was actually studying just one, very specific unit of heredity, which turned out to be DNA. He didn't know that the behavior he was observing and recording was particular to the way that DNA behaved. Nevertheless, his choice to limit his observations to a single category of easily observable hereditary traits has been reflected by the majority of genetics research ever since.

In fact, virtually all the key discoveries in genetics during the 20th century were focused almost exclusively on DNA — and not just on DNA alone, but on the "good" DNA genes that were stable and behaved in ways that made them relatively easy to isolate, observe and work with. Take, for example, the discovery, by Hermann Muller, that he could produce mutations in the fruit fly *Drosophila melanogaster* by bombarding it with X-ray radiation. Every change in an X-rayed fly's phenotype, or appearance — a deformed wing, for instance — would be visible evidence of a gene controlling that trait. Combined with Mendel's selection method, mutating living organisms with radiation dramatically increased the number of genes that could be identified.

DNA is particularly susceptible to radiation damage, so traits that were obviously encoded by DNA genes could be readily flushed out by this technique. But this means the converse is also true: that traits *not* obviously encoded by DNA — the kinds that might, for example, account for some of the differences between mice and men — would be completely overlooked.

≡

There's another piece of enlightening historical context that helps explain why the study of genetics has been so focused on the DNA molecule. According to Michael Rogers, whose book *Biohazard* is the definitive journalistic chronicle of the early days of genetic engineering, the field's DNA focus intensified after a mass influx of renegade physicists in the years surrounding World War II and the development of the atomic bomb. Two of the most prominent physicists of the century — Niels Bohr and Erwin Schrödinger — had suggested that the laws of physics weren't enough to explain the physical basis for life, and that there might be some unknown principles about living organisms that were waiting to be explored.

"For post-Bomb physicists, Schrödinger's suggestion of fresh pastures launched something of an interdisciplinary land rush," wrote Rogers. "By the early 1950s, molecular genetics had been launched with a whole new complement of strayed physicists and chemists, a development that would change the shape of biology to come just as surely as Copernicus changed the direction of astronomy."[6]

The intellectual challenge to these scientists, according to Rogers, was that humans had virtually no control over their heredity. Several of these scientists had worked on the Manhattan Project. Because of work they themselves had done, even nuclear fission — the most destructive physical force ever unleashed — could be controlled by humans to some degree. But our ability to control the replication and expression of DNA molecules was, at that time, beyond our grasp.

Then Watson and Crick gave shape and form to DNA. And in relatively short order, DNA quickly became the "atom" of biology, the *über* molecule, firmly installed at the nexus of virtually all modern biological research. With DNA at its core — easy to study, easy to manipulate — the race to control heredity had begun.

Less than 30 years later, the technology of recombinant DNA was invented. A stunning achievement built on more than half a century of scientific discovery, it was invented in 1973 by two San Francisco Bay Area-based biochemists, Herbert Boyer, at the University of California, San Francisco, and Stanley Cohen, at Stanford University.[7] By redeploying certain powerful chemicals found in living cells to splice DNA from an African clawed toad into the DNA of *E. coli* bacteria, they "genetically engineered" the first transgenic organism. When the *E. coli* bacteria reproduced, the next generation of the bacteria "inherited" and replicated the toad gene right along with its native bacterial DNA.

Some scientists were concerned about the radical nature of this new technique for tinkering with the heredity of living things, but others downplayed its significance. In their view, recombinant DNA was just another victory in humankind's quest to control nature. In fact, this "no big deal" attitude continues to be one of the primary arguments for convincing the public that genetic engineering and the human creation of transgenic organisms are simply the latest steps forward in the continuum of human progress.

It's absolutely true that man has indeed been tinkering with the genes of living things for centuries. Wild plants were domesticated into food crops by the earliest farmers, who selectively bred them for traits (like drought resistance) that would increase their hardiness and yields, and today's sophisticated agricultural techniques still have largely the same goals. Farm animals were domesticated by devising various crossbreeding methods to enhance the traits that would improve their meat, milk and fleece yields. Over the years, breeders have also tailored the appearance, behavior and temperament of all manner of pets — from goldfish to Labradoodles — in accordance with human needs and desires.

But the view that recombinant DNA is just the next step in breeding technology makes too light of Boyer and Cohen's accomplishment. Their invention of the technique that could produce transgenic bacteria represented the first time humans had purposely breached — and bridged — the species barrier. Before recombinant DNA, not even the most sophisticated plant or animal breeding technique could extract the DNA molecules responsible for a specific trait from one species and insert them directly into the genetic material of an entirely unrelated species. Such a feat was impossible without the methods for manipulating DNA that Boyer and Cohen invented.

The question from the perspective of risk is not whether there is anything inherently right or wrong about man's desire to "control" biology or "conquer" nature. We've been designing interventions into natural processes for many hundreds of years. The question is whether biologists' blinkered focus on the ability to manipulate and control a single molecule, DNA, has blinded them to the less obvious aspects of heredity that DNA cannot explain — aspects that in the real world, outside the predictable environment of the laboratory or the production plant, they can neither predict nor control.

In the real world, genes move. Wind and insects shuttle the DNA-laden pollen from one plant to others of its kind and beyond, to varieties in neighboring fields or farther, allowing different populations to interbreed. Even transgenes from commercial crops that are supposed to be unfit for crossbreeding have been found far beyond the fields where they were planted, sometimes permanently lodged in the genomes of neighboring varieties. This very ordinary movement is known as "gene flow," and while it's an important factor in the evolution of plants, it has been considered rare and is not often studied in animals. Yet the work published by several researchers, including the biologist James Mallet at University College London, seems to demonstrate how the boundaries between animal species are porous to gene flow, too.

Even though we don't tend to hear much about it, Mallet says that "greatly improved genetic data" makes it absolutely clear that the movement and incorporation of new genes between animal species is "an ongoing, regular, if not always common feature in nature."[8] Recent outbreaks of West Nile virus in parts of the United States may have resulted from the rise of a new species of hybrid mosquito, for example. And recent research shows that interbreeding between once-distinct varieties of finches on the Galapagos Islands has supercharged their evolution, something that was considered impossible as recently as 1979.[9]

Other crucial natural processes move genes between species as well. Genes, and even entire genomes, can be captured by the cells of wholly different organisms as they feed on, infect or in some other way rub up against each other, by way of what's called horizontal gene transfer. Gene transfer is what causes the spread of antibiotic resistance between different species of bacteria, for example. It's how the DNA that infects wild geese with avian flu is transmitted to chickens, domestic cats, swans, tigers and people. The virus that now produces AIDS was originally called the simian immunodeficiency virus and resided in chimpanzees. Organisms don't need to be very closely related, or even related at all, to exchange genes. In a particularly frightening example from the 1980s, one of the most common and generally benign bacterial species in humans, *Haemophilus influenzae*, somehow acquired the genes necessary to produce a deadly form of meningitis in Brazilian children.[10]

There are also chemicals and substances other than DNA that pass along traits. Stanley Prusiner's prions, as mentioned in the previous chapter, are not DNA at all. A prion is made of protein. It doesn't contain DNA, nor does it shuttle DNA to other organisms like other pathogens do. So it shouldn't be able to reproduce in the way that bacteria or viruses or fungi do. I can get sick from a protein, but I shouldn't be able to be "infected" by one, nor should I be able to pass along that pathogen to any other living things. Infections generally happen only when a pathogen reproduces, and reproduction generally requires DNA (or RNA, in the case of viruses).

But prions reproduce. They undergo a still-mysterious, xerographic kind of self-replication that imprints their shape and infectious properties on normal proteins. The ability to pass along this trait — their shape — means that the trait is heritable. So while a prion is not DNA, it is by definition a gene.

Then there are the networks of molecules that include DNA *and* proteins, in combination with other chemicals, that perform vital regulatory functions in organisms. Some of these networks can change the chemical structure of a protein. Others are on-off "switches" that are responsive to conditions in the cell, like the presence of food or light. Because the information controlling these networks is transmitted to their descendants without being encoded in DNA, they are genes as well.

What's more, the field of behavioral genetics has for several years been challenging the "genes are DNA" assumption with experimental results that consistently fuzz the boundaries between nature and nurture; that is, the predetermined "fate" of DNA versus how organisms respond to the environments in which they live.[11] One remarkable rat study showed that rat pups actually changed the output of their DNA — specifically, the chemicals that affected their ability to handle stress — based on how much maternal nurturing and attention they received.[12] As the Harvard geneticist Richard Lewontin says, this is a basic principle known to all biologists "but ignored by most of them as inconvenient, that the development of an organism is the unique consequence of its genes and the ... environments in which it developed."

≡

Mendel was aware of his bias in terms of the traits he chose to study. He said nothing predictive about things he didn't study, knowing that the subtle traits he left out might not operate by the same rules of heredity as the distinctive ones. Yet today's scientific pronouncements about the safety of transgenic organisms admit no such bias. They've taken Mendel's admittedly biased experimental results about how DNA behaves under tightly controlled conditions and applied them to every living creature on the planet.

What's confounding about the situation is that in the ordinary practice of science, the kind of intensely narrowed focus that Mendel practiced isn't a failing. Instead, it is the modus operandi of the entire scientific enterprise. Scientists base their ideas about cause and effect on evidence they gather as part of an experiment. Mendel did nothing wrong; in fact, he was doing everything right. Experiments have to be rigidly controlled so they consistently produce or cause the same result, over and over again. The ability to reproduce results is the definition of a successful experiment. Any scientific understanding about the mechanisms that make things work, whether at the lab bench or in the field, is dependent on this ability. Results that can be consistently replicated help scientists avoid the dreaded "false positive" result, whereby they declare that an effect exists when in fact it does not.

But this process, which is so fundamental to producing credible science, is not nearly as relevant when the products of experiments move out of the lab and into the field or marketplace, where someone has to decide whether or not they're safe.

"In decision making, which is the eventual purpose of most risk analyses, the goal is not to define the truth of nature," said

Harvey Fineberg, a well-known epidemiologist and risk analyst, once provost of Harvard University and now president of the U.S. Institute of Medicine. "The goal is to use what you can learn about nature to make good decisions. When you have to make a decision, you don't have the luxury of suspending judgment. You can't say, 'The results are inconclusive, let's wait a while, run another experiment, test it again.' *You have to decide.* And how you make that decision is more profound than the complexity of making the truth of nature reveal itself."[13]

In this context, the most difficult problem for scientists to fully comprehend, Fineberg said, is matching what they do to why they are doing it. "Scientists aren't interested in events that they can't predict with perfect fidelity," he said. "They don't measure frequencies in data. They want to get specific results."

What that means for us, who must eat, drink or breathe the fruits of science's labors, is that for the past several decades *decision makers have relied on data from scientists that ignore the purpose of the analysis*: namely, to protect us from undue risk.

So here is the dilemma we face: we persist in asking scientists and engineers to evaluate the risks of scientific innovations based on the experimental data they've been able to reliably record and measure. But even if scientists admit that their attention is focused within the walls of the lab, on the behavior of molecules in a test tube instead of the behavior of organisms in the real world, even if they admit that they're excluding from consideration all the places where there are obvious and critical gaps in what they know — even if they admit they have no data whatsoever on a specific question — that's not enough to trigger scientific precaution. As Roger Brent once said to me, "Unless you can show me the mechanism for risk, it doesn't exist." Period.

Brent's affinity for evidence, not probability, was at the core of his training as an experimental scientist. But risk isn't about what scientists know. It's about what they *don't* know. Risk is about uncertainty. And uncertainty is not what scientists do.

≡

Raise the question of risk with almost any genetics expert from the Watson-Crick-Tilghman school, and in short order the conversation will turn to a meeting that was held more than 30 years ago at the Asilomar Conference Center near Monterey, Calif. In 1975, very shortly after the invention of recombinant DNA by Boyer and Cohen, many of the pioneers of molecular biology came to Asilomar, along with a handful of lawyers and journalists, to discuss whether their experiments using recombinant DNA techniques should be continued.

Scientists themselves, as well as the public, were spooked by the obvious ramifications of this powerful new technology. They were concerned that this technique of recombining DNA from different species — a.k.a. genetic engineering — had the capacity to set loose whole new breeds of unique, and uniquely unstoppable, invasive species upon the planet.

Paul Berg was one of the Asilomar conveners; he later received a Nobel Prize in Chemistry for his early work on recombinant molecules. Berg said the motivation for convening at Asilomar was to see whether those assembled felt they could adequately protect laboratory personnel, the general public and the environment from any hazards that might be directly generated by recombinant DNA experiments.

"In particular, there were speculations that normally innocuous microbes could be changed into human pathogens by introducing genes that rendered them resistant to then available antibiotics, or enabled them to produce dangerous toxins, or transformed them into cancer causing agents," Berg wrote in an essay about the meeting.[14]

The scientists who attended the meeting recommended a self-imposed moratorium on certain kinds of recombinant DNA experiments, specifically those that involved infectious or otherwise dangerous microbes, until researchers felt they'd collected enough experimental evidence to design safety protocols for handling recombinant DNA in the lab environment. These lab protocols were designed, and the U.S. National Institutes of Health formed a Recombinant DNA Advisory Committee (which still exists), to make sure that researchers receiving government money hewed to these protocols. Studies of the recombinant DNA bacteria used in laboratory settings seemed to indicate that they didn't turn into the superbugs that everyone feared, and over time it became a common belief within the field of molecular biology that genetic engineering was safe.

Without this early attention to risk, scientists themselves would have stopped progress in recombinant DNA and the field of molecular biology in their infancy. Instead they've revolutionized more than one global industry, starting with the drug discovery process. By most accounts, this revolution has been a very good thing, in the field of medicine at least. As Roger Brent explained it to me, "Make believe the things that came out of recombinant DNA technology didn't happen. Then, think about what drugs wouldn't have been made."

Without post-1975 recombinant DNA technology, for example, we would not have drugs like Tagamet or Prilosec. We would not have antidepressants. We wouldn't have synthetic human insulin. Without the anti-HIV retroviral drugs like AZT, the development of which also depended on these tools, many thousands more people would have died from HIV/AIDS. A more commonplace but nearly forgotten boon of recombinant DNA was its role nearly 30 years ago in developing more effective detergents that used recombinant enzymes. These new detergents quickly replaced the highly polluting phosphate-based detergents that spurred the excessive growth of algae in freshwater lakes and streams.

The benefits of using recombinant DNA to "manufacture" foreign proteins in a sterile laboratory are clear, as is the utility of the tools that allow drug developers to pinpoint the exact molecule that causes the infection they want to treat. The re-engineered bacteria are "defanged" — denatured of their disease-causing bits — and confined to labs. These organisms are essentially laboratory tools, used to produce an industrial chemical that is then used in the same way as any other. Whether ingested as a drug or dispensed as a detergent or some other product, these products of recombinant DNA can, with relative ease, be contained and evaluated for toxicity and other standard chemical risk factors.

That's not to say that people don't have legitimate concerns about using bacteria or viruses to do this kind of work. Others feel that we're obsessed with pharmaceutical quick fixes and that the billions spent on HIV or cancer drug development, for example, could be better spent on prevention. Some are bothered that these sophisticated products are almost exclusively boons to the developed world. It's hard to argue against any of those points. Progress is in the eye of the beholder, and in any event, it's not risk-free. The science-based concerns for these lab-centric types of products seem to have been dealt with responsibly, however, and those who are lucky or wealthy enough are reaping the benefits. In any case, none of the good that has come of these things — and there has been an enormous amount of good — would have occurred if the use of recombinant DNA in the lab had been banned for fear of being misused.

But the behavior and application of recombinant DNA in the laboratory, which were the focus of the Asilomar meeting, are not the focus of this book. Instead, the subjects here are the products and the byproducts of *transgenesis* — of using recombinant DNA and genetic engineering techniques to create wholly new kinds of living organisms. The transgenic organisms that are of the most pressing concern today don't just sit in a lab vat and produce inert proteins for harvest and resale. These plants and animals and microbes become the *living* products of recombinant DNA. Billions of transgenics have already been released into the marketplace and thus into our food, our water and the air that we breathe, breeding and exchanging their genetic material with each other and with us. At the rate we're going, billions more are on the way.

And because these organisms are alive, and because there are so many of them, and because they are out of our reach and control, and because they have been engineered, and because scientists have paid so little attention to the factors besides DNA that affect the heredity and development of living things, transgenics present dramatically different risks than the well-behaved recombinant DNA that's confined to the laboratory. Genes are far more mobile and their behavior is far less predictable than any risk assessment of transgenics acknowledges. The more transgenics there are, the more transgenics there will be, in more places than anyone had planned or imagined. And once lodged in the right place, transgenes can be passed along to future generations ad infinitum.

CHAPTER 3.
THE EFFECTS OF BIOTECH AT SCALE

"Passed along to future generations ad infinitum" is not a phrase you want to hear about any potential risk, no matter what the scale. But the correlation between risk and scale — in this case, the number of transgenic organisms actually in circulation — may turn out to be a very important assessment, and one that today is not being made. That's because biology is heavily influenced by what are often called threshold effects. A threshold effect is what happens when some type of harm builds up over time and overwhelms the ability of an organism or sometimes an entire ecosystem to recover or heal the damage. Threshold effects are always harmful, and they're often fatal.

Harm usually comes to living things by way of some kind of stress. Stress can be triggered by the toxic cocktail you (or the plants on the side of the road) take in on your daily commute, by what happens to your immune system when you've caught a cold, by too much heat or bright sunlight, or even by the change in your stomach's acid content after a meal. No matter what its genesis, stress sets off a whole series of internal reactions. If the organism can't manage these reactions — if your lungs and blood can't break down and eject the pollutants you inhaled during your commute, for example — then the organism is damaged.

When stress disrupts the function of an organism and the damage can't be repaired, unpredictable changes can take place inside living cells, including the permanent changes to DNA or RNA that are known as mutations. And while mutation is something that happens naturally as part of a normal organism's life, the mutations that occur as a result of stress cause the cell to undergo changes that generally aren't for the good.

Radiation, for example, can have several different kinds of threshold effects. It can trigger genetic aberrations at low doses, sterility as the dosage increases, and, at very high doses, effects on fetuses and early death.[1] This is why X-ray technicians and nuclear plant workers wear tags to monitor the build-up of their exposure. On a larger scale, the havoc that invasive species wreak on the environment, as noted in Chapter 1, is an example of how threshold effects affect not only individual organisms, but entire populations and ecosystems.

Once the damage has gone beyond the threshold, these effects can be cataclysmic. Both the ecosystem and the physiology of the affected individuals can change so fundamentally that they require a new definition of what it means to be "normal." The ecosystem that supported bacteria (that is, most of the planet) was fundamentally different before antibiotic resistance became an epidemic. Because of the enormous quantity we used — i.e., the scale at which we applied antibiotics to the entire ecosystem where bacteria lived (inside us, in hospitals, everywhere) — we pushed the ecosystem past the point where it could heal itself. We can't reverse radiation damage. We can't reverse the effects of invasive species. And we can't restore resistant bacteria to their former state of vulnerability.

"If antibiotic resistance were reversible, we should be able to restore resistant bacteria to complete susceptibility, but we can't," said Jack Heinemann. "That's the threshold effect — once you've crossed it, there's no going back. These resistant bacteria have permanently changed the ecosystem." Heinemann, an American-born-and-educated molecular biologist and associate professor* at the University of Christchurch in New Zealand, is also founder and director of the Centre for Integrative Research on Biosafety in New Zealand. He has spent the last several years working with scientists and scholars from several disciplines to better understand and manage risk and safety issues for emerging biotechnologies, including transgenics.

Consider the possibility that we've created a transgenic plant, said Heinemann, that without knowing it, is just slightly toxic, but not fatally, to some species that's important to us, like a worm or a butterfly. What if the transgene's effects on the worm are so small that we can't detect them by doing ordinary testing — i.e., a sample from a field here, a sample from a worm there? Does that mean the transgene is having no effect? Or does it mean that the toxic effect might be building up over time and at some point will push across the threshold?

At this point, no one knows. Because biotech advocates remain adamant that the products of genetic engineering are not inherently harmful, the concentration and movement of commercial transgenics is neither tracked nor monitored in the field. But with nuclear waste expert Todd La Porte's earlier exhortation in mind — that biotechnology has never been tested at scale — if its advocates are wrong, how big a problem might we be facing? At what scale have transgenics already been released, and at what scale are they being developed?

≡

The most logical place to begin an inventory of transgenics is with plant agriculture. Transgenic crops are by far the most plentiful source of widely available and legally approved transgenic organisms on the planet.

In 2004, according to one annual review of the industry, approved biotech crops covered 81 million hectares,^ up from almost 68 million in 2003 — the second highest increase on record.

---

* This is equivalent to the rank of full professor in the U.S.

^ A hectare is equivalent to 2.47 acres.

That year, transgenic strains accounted for nearly one-third of the global supply of soybeans, maize, cotton and canola. The cumulative numbers are even more impressive. From 1996 to 2004, according to the review, the cumulative global crop area devoted to transgenics added up to 385 million hectares, an area it calculated as equivalent to 40 percent of the total arable land area of the U.S. or China.[2]

Most of these transgenic crops were herbicide-tolerant plants — specifically, soybeans engineered to resist the application of the weed killer glyphosate, best known by the brand name Roundup. In 2004, "Roundup Ready" soybeans accounted for more than half of all soybeans grown in the world, on more than 48 million hectares. In second place but racing to catch up were crops engineered with DNA from *Bacillus thuringiensis* or *Bt,* a soil bacterium that produces a protein toxic to certain kinds of insect pests, particularly worms. To date, *Bt* has been inserted into corn, cotton, potato and tomato plants, with *Bt* canola on the way. Next-generation transgenic cotton and maize are now grown with "stacked" Roundup Ready and *Bt* transgenes. Roundup Ready transgenes also have been deployed in canola, cotton and sugar beets.[3]

While this may seem to be a staggering amount of acreage considering the continued controversies surrounding transgenic food crops, it's important to note that these estimates are low. They account only for approved, official plantings of transgenics. As a result of natural forces like gene flow, transgenes have already moved far beyond the fields where they were planted and have inserted themselves into the genetic material of many traditionally bred crops, as well as wild plants and weeds. No one has tried to count or map the distribution of these "unofficial" transgenic organisms, created by the ongoing, normal mechanisms of reproduction and gene transfer. Even plant material that has decomposed into soil and water runoff can transfer DNA to other plants and organisms. This unofficial movement of transgenes is one of the major differences in the risk profile of recombinant DNA in the lab and transgenic organisms in the field.

Another word to describe "unofficial movement" is *contamination.* The term inspires fear for many people, but it doesn't necessarily have a negative connotation in biology. Contamination is simply the presence of any foreign substance in significant concentrations. Many people, including genetics researchers, use the term "contamination" to describe the unplanned spread of transgenes into other varieties and species.

Genetic contamination is different from the type of contamination that most people have encountered or heard of in the past. A traditional soybean or corn plant that's been contaminated by transgenes isn't like a nuclear power plant contaminated by radiation. There are no genetic Geiger counters that clack a warning in the presence of transgenes; no "clouds" or "plumes" that are released when a transgene infiltrates an organism; no way to hose down a plant or vacuum up its transgenic DNA, or to filter it out of the land or water and be done with it. Detecting DNA requires expensive, delicate machines and skilled technicians to operate them. Even if it were easy to detect, genetic contamination cannot be reversed with today's scientific knowledge; there is no "antidote" that can yank DNA out of a contaminated organism or population.

DNA isn't even deterred by death. Though it's inert when sitting in a test tube, one of DNA's remarkable properties is that genes can "reintegrate" from a dead organism into one that's alive. In fact, the very experiment that proved DNA was genetic material, conducted in the early 1920s, was one in which DNA from dead, lethal bacteria was mixed with live, harmless bacteria — and subsequently killed the mice who were injected with the mixture. Many types of bacteria have these specialized mechanisms for pulling DNA out of the environment and incorporating it into their own genetic material in order to adapt and survive. Other organisms that are not naturally so inclined also can be cajoled into taking up foreign DNA. As we'll see, that ability is the basis for most genetic engineering in the lab.

Meanwhile, in the field, transgenes have already contaminated traditional seed supplies for several important crops in many countries around the world, including corn, soy, cotton, rice, canola and papaya.[4] In each of these circumstances, no one is sure exactly how the contamination happened, whether by gene flow or by mixing seeds together. In mid-2005 it was illegal in China to sell transgenic rice as food; nevertheless, at that time it had already twice been found for sale in Chinese markets. Since then, it's been confirmed that unapproved transgenic rice has been found in U.S. and Indian long grain supplies; the rice market has plummeted as a result. So it seems likely that transgenic rice, too, is already planted in open fields, and there is further speculation that transgenes have already spread to the genomes of the traditional rice supply.[5]

Transgenic wheat has been developed and is ready for commercial release, but intense controversy, both about contaminating the traditional seed supply and about consumer acceptance, has kept it off the market for now. No one expects it's been permanently shelved, however; it was approved long ago by the USDA. In addition, several varieties of squash and banana — including some that are being tested as "carriers" for live vaccines — have been engineered with virus resistance. Virus-resistant transgenic papayas have been widely planted in Hawaii and Thailand despite ongoing protests, replacing conventional varieties; their transgenes have contaminated the remaining conventional crops in both regions.[6,7,8] Many other transgenic food crops are in development as well. And the biotech industry has already started to announce the second generation of transgenic crops, including transgenically decaffeinated coffee,[9] which will be directly targeted to consumers, not just to farmers.

The development of transgenic plants isn't limited to crops. A variety of creeping bentgrass that's widely used on golf courses has been re-engineered with an herbicide resistance gene like the one found in Roundup Ready soy. Developed by The Scotts Company and Monsanto Company, transgenic bentgrass ended up on the front page of *The New York Times* in 2004, when scientists from the EPA found that it had pollinated plants with its transgenes as far as 13 miles downwind.[10]

Also of great environmental concern are the transgenic trees that are already planted in many countries around the world, including the U.S., China, Canada, Belgium and Chile. Transgenic trees are unlike transgenic crops, which can theoretically be harvested every year before their transgenes can spread too far or be exposed to environmental stress (like drought or flood, new diseases or pollutants) that could cause unpredictable genetic changes.[11] Even when trees are engineered as rapid-growth crops for paper or wood, they still live much longer in their ecosystems than food crops. This distinction is important, because trees are fundamental providers of food and shelter for so many other organisms and are an integral part of regional ecologies. Thus transgenic trees are particularly worrisome in terms of both short- and long-term environmental effects.

What might happen to the forests in the eastern U.S., for example, if chestnut trees that have been missing from the ecosystem for 100 years were suddenly to make a comeback — especially if they've been outfitted for survival with disease- or insect-resistant transgenes? One concern is that the transgenic tree would itself become an invasive species, an unpleasantly ironic scenario. Another is that the disease-resistant chemical produced by the transgene might kill or harm other plant life, birds, insects and animals. In cases where trees have been modified for rapid growth and harvest and thus are capable of "using" the soil much more intensely, there are also concerns about soil fertility and health.

That's why ecologists were dismayed to discover that in 2004, *more than a million* transgenic poplar trees were planted all over the Chinese countryside, but no one recorded where. One Chinese scientist reported to a United Nations panel that the transgenic poplars, engineered to resist leaf-eating insects, "are so widely planted in northern China that pollen and seed dispersal cannot be prevented."[12]

Even more worrisome are the transgenic trees modified by a professor at the University of Georgia to take up mercury from contaminated soil. Planted in open field trials in Connecticut and

Alabama, the trees have been modified to express a chemical that converts the toxic metal into one of its less toxic forms and expel it into the air — where, its developers say, it will be "safely" diluted.[13]

But "safe" is an odd term to use in connection with mercury in any form. Mercury is one of the most toxic substances on the planet. The air in a room can reach contamination levels just from the mercury in a broken thermometer. One drop can pollute an entire lake. Mercury accumulates in living tissue in increasingly concentrated levels as it moves up the food chain through plants.[14] Mercury ingested during pregnancy can harm the brain and nervous system of a fetus and can later affect a child's memory, attention and language skills.[15]

It's true that some forms of mercury are more dangerous than others. But in addition to whatever other unknown effects transgenic trees might have on the ecosystem, what might be the demonstrable health or environmental benefit of simply re-depositing a different form of mercury into the air, where it will be recaptured by snow and rainfall, fall back to earth and be consumed in drinking water and absorbed by fish that we will eat?[16] Regardless of the answer to that question, there is a financial benefit, for someone at least. The argument as presented by a local government official is that transgenic trees will allow land condemned because of mercury pollution to be reclaimed for real-estate development.

Field tests in Danbury, Conn., were supported by funding from the EPA, as well as from the biotech company that one of the University of Georgia professors co-founded to commercialize the remediation technology. Although the field tests were to run through the 2004 growing season, by July 2005, the experiment apparently was still in its trial phase; the professor didn't respond to several attempts to contact him for an update.[17]

≡

While they are nowhere near as widespread as transgenic plants, the number of transgenic animal species that populate the biotech world is growing quickly as well.

In the lead, not surprisingly, are transgenic mice. More than 95 percent of all the transgenic animals in the world are mice, mostly used for lab experiments. As noted earlier, their genetic structure is so similar to ours that scientists have turned the mouse genome into a kind of Tinkertoy, "inventing" (and patenting) new mouse varieties with regularity by snapping genes in and out, for basic research as well as to test various types of human drugs and therapies.

While the genetic structure of mice is the attraction for medical researchers, it's the physiology of pigs that they find intriguing. As it turns out, the organs of domestic pigs are about the same size as those of humans, and are "plumbed" in the same general way, which makes them strong candidates for organ transplants to humans. In order to keep our immune systems from rejecting a pig organ transplant, scientists have created transgenic pigs whose genomes lack the specific bit of DNA that causes the rejection reaction. There are some significant risks associated with this process, including the unknown probability that dormant pig and human retroviruses could recombine into a new and unstoppable disease; nevertheless, scientists in South Korea are racing to be first to commercialize the technology.

Transgenic fish, as well as poultry, swine, goats, cattle and other livestock, are being modified so their cells can be used as "factories" for pharmaceutical and other products, including human blood-clotting factor, and also as models for human disease. Non-allergenic transgenic cats are already on the market. Insects are being modified to produce recombinant proteins like antibodies and enzymes, and the genes that allow them to harbor infectious diseases are being "deprogrammed." Also, there's the famous Glo-Fish, the first commercially available transgenic animal, genetically engineered to emit a phosphorescent glow for no beneficial reason whatsoever.

One of the more troubling issues when we consider questions of scale comes from the fact that transgenic microbes and viruses are the focus of much commercial development. Adenoviruses, a type of virus that generally causes respiratory tract and eye infections, have been engineered to selectively burst cancer cells,[18] raising concerns about what other kinds of cells these common viruses could be engineered to "selectively" burst — or whether they would quickly mutate to burst other kinds of cells as well. Some studies have shown that viruses become more infective in "naïve" host populations that have no resistance to them. Thus releasing genetically engineered viruses into the environment might result in their becoming successfully established in organisms other than the target.[19]

Many strains of bacteria are being genetically modified for commercial use as well. The bacterium that's responsible for tooth decay has now been engineered to prevent it.[20] Vaginal bacteria have been genetically altered to create a "living condom" packed with unfriendly proteins that its developers hope will be able to prevent HIV infection.[21]

Most notable in this category — for sheer ambition, at least — is the stated intention of Craig Venter (quoted in the introduction regarding how little we know about biology) to sequence the genomes of the microbial world. In 2004, he received $8.95 million for two years to sequence 130 marine microbes. Then in March 2005, he received another $2.5 million from the Sloan Foundation to sequence the microbes in New York City's air. Venter's goal is to get enough information "to create microbes from scratch that can produce clean energy or curb global warming."[22]

Sequencing the genomes of microbes is a fine idea. There's much to be learned from these ubiquitous and vital organisms. But with all due respect, trying to create entirely new ones and cut them loose in the environment seems like the epitome of playing with fire. If the scientific elite like Venter and Princeton's Tilghman already admit to not knowing much about the genetics of comparatively big, complex organisms like mammals, what they don't know about microbes — not least of which is how to control them — is nearly infinite in comparison. How could that be, when microbes are so small and so simple, relative to us? Here's what Joshua Lederberg, who received a Nobel Prize for his pioneering work on bacterial genetics, had to say on the subject:

> What makes microbial evolution particularly intriguing, and worrisome, is a combination of vast populations and intense fluctuations in those populations. It's a formula for top-speed evolution. Microbial populations may fluctuate by factors of 10 billion on a daily cycle as they move between hosts, or as they encounter antibiotics, antibodies, or other natural hazards. A simple comparison of the pace of evolution between microbes and their multicellular hosts suggests a millionfold or billionfold advantage to the microbe. A year in the life of bacteria would easily match the span of mammalian evolution! By that metric, we would seem to be playing out of our evolutionary league.[23]

Yet among the scores of articles I've read about Venter's transgenic microbe projects, there was not one mention of these known scientific realities, or of risk.

In addition, human gene therapies and other interventions that use live transgenic organisms, including transgenic vaccines, are already well along in trial phases — despite serious setbacks and concerns, including fatalities. Gene therapy saved the lives of French twins with damaged immune systems, for example, but it was later found that the virus carrying the therapeutic genes inserted itself next to a cancer-causing gene, activating it and causing leukemia in both boys.[24] In the U.S., 18-year-old Jesse Gelsinger, who suffered from a rare liver disorder, died as a result of a severe immune reaction to a gene therapy experiment at the University of Pennsylvania, where he was injected with transgenic viruses designed to carry healthy copies of a gene into his body.[25]

≡

What are some other concerns about transgenic microbes? We can look back to the first patented organism for a clue. Back in

1980, Ananda Chakrabarty was awarded the first patent on a living organism — an altered *Pseudomonas* bacterium that was capable of breaking down crude oil into stuff that could serve as "food" for aquatic life.[26] A domesticated bacterium with this kind of genetic equipment, it was thought, would be a great boon — for the same reason our University of Georgia professor invented his mercury-aspirating tree. The concept is known as "bioremediation," and in practice it would set loose living organisms to clean up the messes we've made in the environment.

Chakrabarty's microbe was supposed to eat up oil spills. But it never hit the market. Why? It's not exactly clear. All I could find out was that these carefully designed oil-eating microbes quickly evolved, precisely as Joshua Lederberg might have predicted. As Chakrabarty himself delicately phrased it in a 2002 interview, "The bacteria by itself is non-toxic but once in the open environment it can combine with pathogenic elements and show undesirable results."[27]

Let us keep in mind that the "open environment" under discussion is ocean water, which covers about 70 percent of the planet. Yes, that might present a problem.

CHAPTER 4.

THE RISKS OF GOING NATIVE

One way to describe what might have happened to Chakrabarty's bacteria is that they became feral. They stopped being domesticated bacteria, as they were trained to be, and started acting like wild bacteria again — rapidly evolving, opportunistic, infinitely hungry bacteria

Science-based concerns about transgenics "going native" in this way were a key topic in the 2002 *Animal Biotechnology* study that was published by the U.S. National Academies. The National Academies is an elite private membership organization, closely allied with the federal government, which represents the most distinguished researchers in the country. In order to provide advice to the government, it brings together committees of experts from all areas of science and technology that produce peer-reviewed studies and reports on a wide variety of critical scientific issues of the day. Over the course of the past decade or so, it has published five reports that have focused on various safety issues and other potential effects of transgenic organisms.

Members of the *Animal Biotechnology* study committee were most adamant about the hazards presented by transgenic insects and aquatic organisms, primarily because their mobility makes them almost impossible to contain. Unlike domestic farm birds and mammals, both insects and fish, according to the report, "can easily become feral and compete with indigenous populations."[1]

A case in point: Transgenic salmon, already under development, could create several very difficult-to-manage situations should they break loose from their farm pens (which everyone assumes they will). Because fish can so quickly revert to their feral state, and because wild salmon already are known to mate with species of trout, contamination of the wild trout by salmon transgenes is highly probable — to highly uncertain effect. The *Animal Biotechnology* report noted that non-transgenic cultivated salmon that have escaped into the wild already pose ecological and genetic risks to native stocks. But the transgenic salmon presently awaiting regulatory approval could be an even bigger threat. Salmon engineered with a fast-growth gene are growing four to six times faster than wild salmon in the lab. They are also 20 percent more efficient at converting food to body weight than their wild brethren, and they eat faster, too. This suggests that the transgenic salmon would out-compete the wild type — i.e., become an invasive species — a possibility that is of "immediate concern," according to the study.[2]

Yet even in the face of such authoritative warnings, transgenic salmon with a fast-growth gene are already in the queue for regulatory approval, and similarly engineered trout are on the way as well.

Some believe that designing sterile salmon could alleviate many of the concerns that arise from fish escaping. But this approach works only if the sterilization process is 100 percent effective — not just in the controlled development environment, but in a field-tested, real-life context where none of the usual laboratory controls apply: a dicey proposition at best. There simply is no data to support that the approach will remain effective when (not "if," please note) the fish do escape.

Another important issue about transgenic fish was raised by Ronald Coleman, a fish expert who teaches at California State University in Sacramento.

While many people don't like the idea of genetic engineering because of the "yuck" factor of putting DNA from one species into the DNA of something entirely different, Coleman noted that there seems to be less concern about supercharging a species with genes from closer relatives.[3]

In the commercial fish industry, for example, there has been a great deal of interest in developing methods that would allow a popular eating fish like the trout to survive and thrive in colder water. If the fish could keep growing through the winter, the reasoning goes, they could get to a marketable size and weight sooner. Scientists long ago isolated a sequence of DNA, found in various forms in various species of fish, that provides this "antifreeze" quality. In fact, the first transgenic tomato had been (unsuccessfully) re-engineered with an antifreeze gene from Arctic flounder, eliciting the proto-"yuck" response.[4] (This was a slight overreaction, as the sequence was not literally plucked from a fish but, as is true with most commercial transgenes, was instead synthesized from the chemical composition of the flounder's DNA and optimized to work in a plant's genome.)

Yuck factor aside, Coleman said he tells his students that "the systems in an organism are highly integrated and when you fiddle with one thing, you will likely alter a bunch of other co-adapted systems, and so the net result may not be what you were expecting."

For example, he noted, "many of the fishes that have antifreeze genes have a bunch of adaptations that go along with them. For one, they have a different kidney. Why? Because if they had the regular fish kidney it would constantly filter out the large antifreeze protein molecules from the blood. So in concert with evolving the ability to produce antifreeze, they had to evolve a different kidney. Trout didn't evolve antifreeze and they didn't evolve a different kidney. So even if you stuck the antifreeze gene in the trout, what you might eventually find is that rather than a bunch of fast-growing fish, you have a bunch of weird fish with kidney issues, or fish that don't grow faster at all because they are consuming tons of energy making these antifreeze molecules (which are 'expensive' to make) only to immediately filter them out of the blood."

These are just two scenarios of unexpected consequences that come from knowing "just a little about fish kidneys," said Coleman. "Imagine all the possible unexpected outcomes from all the systems in a body we know so little about (which is most of them)."[5]

Perhaps even more troublesome in the context of going feral is the introduction of transgenic insects. The reasons scientists decide to pursue insects as a target for transgenesis are more varied than for fish, which are generally altered to improve their value as a food source. In contrast, the most promising and the most threatening insects that are being genetically modified are those that have been altered to either control or eliminate their roles as food pests, or to reduce transmission rates of the diseases they carry. Mosquitoes that carry malaria are at the top of this list, followed closely by the "kissing bugs" that carry the devastating Chagas disease that can be passed along from human parent to child. Tsetse flies, Mediterranean fruit flies and black flies are also targets in this category, as are the moths of pink bollworms, which threaten cotton crops.

In addition to transgenic insects designed to control human or plant disease, honeybees are being engineered for resistance to pesticides, diseases and parasites; silkworms are producing drugs and industrial proteins; and beetles and weevils are altered to eat only one kind of noxious weed, turning them into a kind of living herbicide. (Good thing they can't swim, or someone no doubt already would have engineered them to eat the noxious *Caulerpa* that has invaded the world's oceans.)

No one knows whether transgenic insects will have a lasting effect once released. The "success" of some genetically altered insects will depend on whether fertile transgenic varieties could replace wild insect populations and become established in the environment. Replacing wild populations is a much dicier proposition, safety-wise — especially in the case of the biting and sucking insects. Because they exchange so much DNA with their environment and their food sources (a.k.a. "us"), the potential is much greater that their transgenic traits will spread throughout the insect population, potentially making pre-existing pest problems worse or creating altogether new challenges. In the case of transgenic honeybees, modifying their genetic composition could alter the composition of their honey, creating a potential food-safety concern for the humans or animals that eat it.[6]

Let's consider more closely the high probability of genetic exchange between wild and transgenic insects: It's possible that transgenic insects released to control the spread of disease could actually have the unintended consequence of enabling an insect to more effectively spread disease, or even carry a human disease or human-specific genetic alteration it was never before able to transmit.

For example, in 2004, researchers at the University of California, Riverside, found that their transgenic mosquitoes, altered so that they could not carry or pass along malaria, were less fertile and less healthy than their wild counterparts.[7] This was considered a problem, because the mosquitoes wouldn't be able to survive in the wild. And so the question these researchers are now asking is how to "engineer" these already transgenic insects to be invasive enough that they can overwhelm the existing wild population. In other words, in mid-2005, experiments were actually under way *to purposely design an invasive species.*[8]

What might be the effect if these newly super-fit transgenic mosquitoes breed with other, similar insects — and by gene flow pass along the transgenes outside the mosquito population? Or if after all that trouble, one of these similar insects transfers back to the super-fit population the ability to carry malaria? Or if a microbe as dangerous as or worse than malaria learns to colonize the new mosquito?

These are pretty scary scenarios. Because there's ample evidence that gene flow is common between insect species, many researchers and government regulators are rightfully worried. But given the many potential applications for transgenic insects, research continues unabated, as does commercial development of virtually all animal species — even though "[i]n only a few cases was it possible to state that an issue brought to our table was not of concern," according to John Vandenbergh, a professor of zoology at North Carolina State University who chaired the *Animal Biotechnology* study committee. "Much of the basic biology underlying the techniques remains to be discovered, and we have only partial information on the consequences of using biotechnological techniques."

≡

Some scientists say that the ability to contain or confine transgenic organisms would "solve" at least the most obvious of these problems. However, by one count, by mid-2005 some 27 countries had already reported a total of 62 cases of contamination of food, animal feed, seeds or wild plants with transgenes.[9] But even more disturbing than the problem of containment is one of the proposed solutions: the *biological*, rather than physical, confinement of transgenics that could potentially become invasive species.[10]

When I first saw the term "biological confinement," it took me a minute to parse what it meant. What it assumes from the start is that (a) we can no longer *physically* confine transgenic organisms (i.e., crops to their own fields, salmon to their pens), and (b) that it is pointless to even attempt physical confinement.

That leaves us with biological methods, like genetically induced sterility or genetically engineering fish to rely on a man-made substance for survival — like Skittles or beef jerky, maybe — so they'd die if they escaped into the wild and couldn't quite make it upstream to the 7-Eleven. For engineered microbes, one idea is to make them such voracious energy or food hogs that they can't compete with native bacteria or fungi. Given Joshua Lederberg's earlier statement about the ultra-rapid adaptability of the microbial world, I have to admit that these "solutions" make the hair stand up on the back of my neck.

Another proposal is to use a chemical to trigger "suicide" genes if microbes escape confinement, although no one has ever tested this method. A similar commercial technology, co-developed by the USDA and a seed industry partner, Delta & Pine Land Company (now owned by Monsanto), genetically modifies plants to produce sterile seeds, eliminating the possibility that farmers would reuse harvested seed. Critics fear that these plants would irreversibly spread their sterility to non-transgenic crops and across species to other plants by contamination. The technology has been widely condemned as too risky for commercial release, not to mention a serious threat to global food security. But it is far from being abandoned. The USDA and Monsanto both have continued development.

Concerned scientists already consider several kinds of transgenic plants to be dangerous enough to require confinement. One category, ironically, is plants that tolerate herbicides. The concern is that, by gene flow and other natural mechanisms, these plants could then pass on their genetic resistance to plants we actually *want* to kill with herbicides. However, with millions of acres of Roundup Ready soybeans and other herbicide-tolerant plants planted and harvested around the world over the past decade, these transgenes already have long since been discovered in other, non-transgenic varieties — and, in the case of transgenic bentgrass, in entirely different species as well.

Another candidate for confinement would be any kind of organism with engineered traits that allow it to grow faster, reproduce more quickly or vigorously, and live in new types of habitats, like Chakrabarty's oil-munching microbe. This group would also include transgenic fish or shellfish, as well as transgenic insects, all capable of mating with wild counterparts or outcompeting them for food.

Finally, there are the plants and animals engineered to produce pharmaceuticals, vaccines or industrial chemicals — a genre often referred to as "pharming" — which have the capacity to harm people or other species that might accidentally consume them.

≡

Pharming raises some of the more exotic yet obvious risk questions in the world of transgenics. Like with recombinant DNA in bacteria in the early days, the purpose of pharming is simply to use the plant or animal as a cheaper or more productive (or both) living factory for the substance, which will then be harvested.

Pharming is the biological equivalent of offshoring. If you've decided that spider silk would make a great new high-tensile bulletproof fabric, you'll need to get your hands on a serious volume of spiderwebs. But you certainly aren't going to make a market urging a few billion spiders to spin more, faster. You've got to have higher volume than a spider can give you and a renewable and reliable resource that doesn't die as quickly or as easily. Milk produced by farm animals is a well-known renewable resource. Thus you design a goat that can lactate silk, and figure out how to "harvest" the silk from the milk. Theoretically, at least, you've got your renewable resource.

That's one approach to pharming. Another is to create an organism — let's say a transgenic banana tree or corn plant — that expresses a chemical compound that's intended to be eaten in its whole-plant form, as a drug or a vaccine. Here's where pharming represents a different kind of physical hazard to health and the environment. The risks of regular old transgenic corn that we've considered thus far, risks that aren't exactly dismissible, are compounded when you're talking about transgenic corn that's packing some kind of live vaccine or pharmaceutical compound in every one of its kernels.

One scientist concerned with some of the unintended consequences that might result is Norman Ellstrand, a professor in the Department of Genetics at the University of California, Riverside, and director of its Biotechnology Impacts Center. Ellstrand's expertise in the area of gene flow, especially as it relates to the unintentional spread of transgenic plants, has earned him much respect as both a scientist and a fair and credible critic of various aspects of agricultural biotechnology. In 2003, Ellstrand published an essay that speculates about what might happen should a pharm crop "escape" from a limited plot of land here in the U.S. and make its way to another part of the world.

Two issues make Ellstrand's scenario more worrisome than our existing slate of risk concerns. One is that the drug or chemical compound produced by the pharm crop is harmless if consumed in low concentrations, but triggers a threshold effect if enough is consumed over time. For example, what would happen if people or animals were unwittingly eating live vaccines or other pharmaceuticals in their bananas or corn every day, over the course of several years? If the compounds were harmless at low "doses," it's very unlikely someone would detect the compound before it triggered the threshold effect.

The other is not about health, but about human behavior, an element of risk that's often and foolishly disregarded in official assessments. Seed moves easily and often across and beyond borders. By fair means or foul, accidentally or deliberately, seed can easily end up in communities far away from its home. And those various communities may have very different customs, practices and even laws.

In the U.S., for example, big corporations are responsible for developing and selling most agricultural products. As a result, most of the commercial seed in the U.S. is "terminal." That is, farmers don't actually buy seed, they "license" it for a single season. Thus they are obligated by law not to reuse last year's seed, to "terminate" it and buy a new batch each year, rather than save some from the harvest to replant the next season. But most of the world's farmers do what farmers have done since the beginning of time: they save seed. They replant it. They exchange seed with each other and experiment with seed they get from distant sources, trying to improve the health and productivity of their crops. Their food crops are often open-pollinated landraces — varieties of crops that have been developed and improved by the farmers themselves, adapted to local environments and uses — not hybrids developed in laboratories.

So let's suppose that one or a few seeds bearing Ellstrand's "pharmgene" make it to one of these landrace fields in a different country, and they really like their new home. They adapt so successfully that the resulting plants end up producing more pollen, making more viable seed after being pollinated or surviving better than plants without the pharmgene. In other words, they become an invasive species.

When farmers have no legal or cultural imperative to "terminate" the use of a seed, the alien trait can increase in frequency, undetected, generation after generation. This means the compound itself also increases in frequency, as does its concentration in the food supply, until it eventually begins to cause serious health effects in the humans who consume it.

"Seem far-fetched?" wrote Ellstrand. "Each of the components of [this] scenario has a very low probability of happening. And yet, each of the steps is represented by real phenomena." Even though Ellstrand considers that the probability of such an event is low, the gravity of the hazard posed by his scenario is such that "now is the time to start looking for even more effective procedures at containment."[11]

It's not as though something similar hasn't already happened. In 2001, two studies showed that transgenes had contaminated native corn varieties in Oaxaca, Mexico, despite a national multiyear moratorium on growing transgenics. This, according to Ellstrand, "likely represent[ed] the migration of those genes across international boundaries."[12]

What to do? Well, just as we saw with other man-made invasive species, scientists are cooking up technologies they say will limit the spread of toxic pharmgenes to what Ellstrand calls the "food and feed streams." You can sterilize the males, vary the number of chromosomes so that the plant cannot produce seeds, insert genetic constructs that sterilize the seeds (the "terminator" technology mentioned earlier), and so on. But to be effective, these techniques simply cannot "leak" — that is, they have to work consistently. And they do not. "I am not aware that any of [these bioconfinement techniques] have been tested thoroughly enough to demonstrate their efficacy in a variety of environments," wrote Ellstrand.[13]

I do feel compelled to point out something here. "Bioconfinement" *is a whole new field of biotechnology, made necessary by the fact that we didn't know how to control the last new field of biotechnology.*

That realization made the 2004 National Academies study on the subject of biological confinement one of the most depressing things I've read in a long time, because the study committee (of which Ellstrand was a member) made it perfectly clear that (a) there *are* no viable, tested alternatives to bioconfinement, and (b) none of the bioconfinement methods we've got are particularly effective, so we probably shouldn't trust them anyhow. The committee suggested using more than one method "to lower the chance of failure." Personally, I do not sleep more soundly based on that suggestion.

Transgenic bentgrass that spreads at a truly alarming rate is one thing. But pharming has even more disturbing connotations, in part because so much of the product development is company-secret. Under present laws, biotech executives cannot be compelled to divulge that information to the public, no matter what drug or compound is being grown in the pharmed plants or animals.

However, a company called ProdiGene, based in Texas, was forced out of the closet in 2002 when it ended up on the front page of newspapers all around the world. The experimental pharm corn it had planted in two open fields in Nebraska and Iowa, which had been genetically altered to grow drugs to prevent diarrhea in pigs, contaminated a soybean and a corn crop. No one knew until after the contamination was made public what exactly ProdiGene was growing; it took the scandal to force the company to divulge what the corn had been engineered to produce. In Iowa, 155 acres of nearby corn were uprooted and burned, and in Nebraska, 500,000 bushels of soybeans had to be destroyed.[14]

And yet research focused on commercial pharming continues apace. Other companies are working on projects like transgenic tobacco that grows edible vaccines, rapeseed plants that grow industrial oils and fish that produce human blood-clotting factor.

But as with all the other examples cited, so far there has been no public consideration of the health and environmental issues that Ellstrand and other scientists have raised.

≡

Basically, there is no good news here. It's hard to imagine how anyone could walk away from these peer-reviewed reports — published by the most elite scientific advisory body in the U.S., and based on research from some of the field's most credible scientists — and still believe that the ongoing deployment of commercial transgenics is without risk. Yet development in all of these areas continues unchecked.

Given the scale at which they've been deployed, the alacrity with which they are being developed, the dearth of knowledge about their long-term behavior or effects and the practical impossibility of confining them, any problem that may present itself as a result of transgenic organisms is likely to be a big one. For those willing to look at the limitations of our knowledge and the nature of technological risk, it's clear that we are brewing a concoction of unknown effects using the genetic material of the entire planet as our stirpot. As industry and regulatory officials continue to ignore or deny the gravity of the situation, the probability increases that a confluence of circumstance and unintended consequences will escalate what might have been a manageable risk to proportions that spiral out of control.

And there is not, nor will there ever be, a traditional risk analysis that on its own can bail us out.

CHAPTER 5.
WHAT GETS MEASURED IS WHAT MATTERS

Used properly and under the right circumstances, traditional risk calculations have proven to be a powerful tool to forecast the odds that a future hazard will come to pass. Over the course of the past several decades alone, for example, risk analysts have been increasingly able to assess the chances of whether and how we will be affected by certain types of hazards presented by life in a modern technological society. With varying degrees of accuracy, they have ways to calculate the probability that we will get sick from the air we breathe, the water we drink, or the additives and pesticides in and on the foods we eat. They can calculate our exposure to risks presented by the drugs we take, the coating on our cookware, the safety glass in our windshields, the performance of our mutual funds and the abrasive compound in our toothpaste. They've calculated the number of years that regular exercise adds to our lives and how many years obesity and smoking subtract from them — as well as the odds that a plane crash or drunk driver will kill us first.

The circumstances in which probability calculations are most likely to be useful are those for which there are lots of recorded data available, based on past experience with similar events. For example, there are many well-known probability distributions for defect rates in manufacturing. But for any circumstance where there isn't a lot of recorded data — that is to say, most other circumstances — forecasting risk using probability math becomes a much more subjective proposition. That's why for the past 20 years or so, risk experts have been sending up signal flares in the academic press, as well as in studies prepared for government regulators, that traditional risk calculations can often be useless in forecasting the likelihood of brand-new hazards, like the ones that very often result from innovations in science and technology.

The main reason is obvious, if you think about it. New technologies don't *have* a history; that's why they're called "new." There isn't enough historical data, if any at all, to calculate risk. The trajectory of the science is unknown. What we will learn about these innovations over time, and in what domains, is uncertain.

In fact, risk analysts have long agreed that the most intractable problem in assessing the risks of innovations is not what we know about them. It's the sparseness of scientific knowledge, and the resulting uncertainty about long-term and unintended consequences. For this reason, over the course of the past 30 years, analysts have been trying to rattle the cages of government regulators and industry leaders, to alert them that traditional methods for evaluating technological risks can be deeply and sometimes even fatally flawed in situations like the ones we face with the products of genetic engineering. It's a quintessential example of where data is scarce and uncertainty is rampant.

≡

Traditional methods of calculating risks are prone to both ask and answer the wrong questions, according to Paul Thurman, a popular professor of data analysis who, among his many appointments, teaches in the Masters of Business Administration programs at Columbia University, the University of California, Berkeley, and the London School of Business.

"If I've built a model based on certain assumptions, that's what I believe," Thurman said. One noteworthy and familiar example was the financial debacle triggered by the hedge fund Long Term Capital Management. The company started with $3 billion in investment capital — and financial models from two Nobel Prize-winning economists — then folded after losing $4.6 billion in less than four months, nearly tanking the global economy in the process. "Their risk models were built around the assumption that market volatilities and stock prices would stay within historical bounds and follow certain patterns or distributions. When they built their models, they didn't include crash events or outliers, like the 1987 crash. What eventually brought them down was the fact they had built these models on relatively stable economies — then out of the blue, banks started busting from Japan to Indonesia, Russia defaulted on its debts, and so did South America. Outlier after outlier tanked their model."[1]

According to Thurman, with whom I discussed these issues at length, this is a typical example — albeit an extreme one — of business as usual in the world of industrial risk assessment. "Many researchers simply believe the numbers that come out of the computer," he said. "They say, 'I have the model; that's the right thing to believe. If there's anything out of pattern, then I must have done something wrong.' They rarely think that the model itself could be wrong. People don't do the simple sniff tests anymore to see if the data makes sense in context — they immediately cut to a quantitative explanation."

What's worse, he said, is that numerical models "infer a scientific precision that isn't there," yet they are accepted at face value as objective data for virtually all risk decisions. "It's almost a binary thing. You're either told by mathematicians that your opinions aren't worthy because you base your arguments on intuition, or you have a PhD in statistics and you can espouse anything you want. But obviously even Nobel winners can build erroneous models."

The probability models that risk analysts use and the way they interpret their data can affect far more than just financial markets. What we decide based on their conclusions can have a life-or-death effect on our daily lives.

For example, there are few behaviors more ordinary than getting in a car and buckling up. Seat belts are widely believed to be the most effective safety devices in vehicles today, reducing by at least 50 percent the risk of being killed or injured in a car accident. Laws in virtually every civilized nation in the world require them. But contrary to popular opinion, seat belts do not, apparently, benefit everyone. In fact, seat belts and the laws that enforce their use are a vivid object lesson for what happens when we accept probability estimates without first asking for or understanding the context in which they're presented.

John Adams, a professor at University College London and author of the book *Risk,*[2] made a splash on the public safety scene some 20 years ago by challenging the statistics presented by safety researchers who were rigorous advocates of seat belt laws.

From the earliest days of the automobile, it has been noted that people change their behavior when they perceive changes in risks to their personal safety. This change in behavior is known as "risk compensation." Adams argues (somewhat controversially, I should note) that we all come equipped with "risk thermostats" that, depending on what changes occur in our environments, inspire us to increase or decrease our reckless behavior to the risk level we're comfortable with. In this case, when people feel safer in their cars, they compensate by driving faster and more recklessly. Governments and industry are complicit as well. Car companies design cars for greater maneuverability at higher speeds, and governments raise the speed limits on highways.

Risk compensation, wrote Adams, seems to be common sense to most people. "About the only area where [the idea of risk compensation] still meets resistance is in the work of people with a professional interest in safety," Adams wrote. "The strength of conviction about what [seat belt] legislation has achieved is remarkably independent of objective evidence."

For example, he said, seat belt advocates rely on two types of research: experiments using crash test dummies and hospital-based reporting. But experiments using crash test dummies do not account for the effect of seat belt-wearing on driving behavior. While crash test studies show that seat belts do absolutely save the lives of people inside the crashed vehicle, they disregard the effects on people *outside the car.* Even inside cars, extending the compulsory seat belt law to children resulted in a 10 percent increase in fatalities and a 12 percent increase in injuries to children. One co-author of a study of children involved in car crashes has said publicly that "the early graduation of kids into adult lap and shoulder belts is a leading cause of child-occupant injuries and deaths."[3]

Yet the belief that seat belts are fully synonymous with safety persists. At the time the British seat belt law was passed in 1981, Adams wrote, none of the law's supporters "appeared to be aware of ... the possibility that there might be a behavioral response to the compulsory wearing" of seat belts. This possibility, he said, had not been investigated in any of the studies the law's supporters cited as evidence.

The moral of this story is not, "Numbers lie; we can't believe them." Instead, it illustrates the power of numbers and statistics in today's culture, and how much we need to respect their influence when wielded by either the naïve or the conniving. We all tend to *believe* "the number," any number, especially if it purports to be the result of a sophisticated mathematical calculation — even when a number is meaningless or misleading in the context of the problem.

≡

Ah, mathematics. The dreaded "M" word. Most people, it seems, at least in the United States, are not math-literate. This particular form of illiteracy — known as innumeracy, the inability to understand or practically use most mathematical concepts — is the ostensible reason most scientists and technical people won't engage in a serious discussion about risk with those outside their numerate tribes. They declare that we, the innumerate public, lack the mental capacity to understand what they, the experts, do; as a result, there can be no common ground for understanding between those who create risk and we who must bear it.

This epidemic of innumeracy is, they claim, the root cause for many of society's ills. The problem with innumerates isn't only that we lack perspective about numbers. Apparently, we also have an exaggerated appreciation for meaningless coincidence. And we believe in pseudoscience. What's more, we tend to personalize things — to be misled by our own experiences rather than being objective and rational and informed by the facts, an error in judgment that apparently no mathematician or scientist ever makes.

Particularly problematic, we are told, is the gap in our understanding between the expert assessment of risks and the naïve way we perceive them. This gap, wrote John Allen Paulos in his book *Innumeracy*, "threatens eventually to lead either to unfounded and crippling anxieties or to impossible and economically paralyzing demands for risk-free guarantees."[4]

Those of us who break out in a sweat in the presence of any math-related artifact more sophisticated than a pie chart may feel there must be truth to Paulos' assertion. But in fact, there is no evidence for it in the risk research. Just one example is the 1979 survey by the chemical giant E. I. du Pont de Nemours and Company (a.k.a. DuPont), conducted while public distrust of big chemical companies was escalating. When asked whether or not they wanted their products to be risk-free, only a small fraction of the respondents said yes.[5] Most people responded that they liked a zero-risk proposition if they could get it (who wouldn't?), but certainly it wasn't a "demand." To conclude that "innumerate" people cannot understand risk because they don't understand sophisticated mathematical concepts is inaccurate at best, and it certainly isn't an *objective* truth.

In fact, it's a rather unsophisticated representation of what risk actually is. For example, a 2001 review of a series of public opinion surveys about biotechnology showed that there is no single "public" that either is anxious or demanding about scientific interventions. Individuals hold very different views about biotechnology, and they have strong opinions about how the industry and its products are managed. They know they have limited knowledge on the subject, and they respond to evidence.[6]

Let's look at a specific incident where this purported gap in public understanding of biotech risk was addressed head-on. In 2003, the U.K. government shelled out £500,000 to address the "ignorance" gap between scientists and laypeople, with a public educational forum on transgenic food called "GM Nation? The Public Debate." (GM stands for "genetically modified.") For three months, broad public debate ensued. But more interesting for our purposes were the so-called Narrow-But-Deep participants. These people were selected at random from the U.K. population to serve as a check that participants in the open debates weren't just those who already knew about the topic, potentially slanting the debate and the results.

If you accept the results of the study, this concern was warranted. The U.K. public as represented by the media, at least, continues to be quite vocal in its lack of support for genetic engineering, and this attitude was, in fact, reflected in the strongly anti-GM tone of the public debates. But over and over again, the Narrow-But-Deep participants — who were asked in-depth questions and provided with more specific scientific information about genetic engineering — were more nuanced in their reactions.

The Narrow-But-Deep responses suggested that when people in the general population are first asked about genetic engineering and transgenic products, they're a little shy and vague about answering, in part because of their lack of knowledge about the subject. But as they become more engaged, they ask more questions and learn more about the issues, and their attitudes take more solid form. Ultimately, while they became more willing to accept some potential benefits from genetic engineering (especially medical benefits and other promised advantages for developing countries), they were also more concerned and uneasy about risks. "In particular," says the report, "the more they choose to discover about GM the more convinced they are that no one knows enough about the long-term effects of GM on human health."[7] While they were in favor of proceeding with caution, the final report stated, participants thought "GM crop technology should not go ahead without further trials and tests, firm regulation, demonstrated benefits to society (not just for producers) and, above all, clear and trusted answers to unresolved questions about health and the environment."[8]

These concerns don't sound like particularly "unfounded and crippling anxieties" or "impossible and economically paralyzing demands." In fact, it seems some of the participants' anxieties about political issues were actually quite well founded. The report noted they had maintained "a strong and wide degree of suspicion about the motives, intentions and behavior" of those making decisions about the products of genetic engineering — especially government and multinational companies. "Such ... lack of trust expresses itself through several avenues," the report said. "One is the suspicion that the government has already taken a decision about GM: the debate was only a camouflage and its results would be ignored."[9]

Sure enough, five months after the findings of the public debate were published, cabinet committee papers were leaked to the newspaper *The Guardian* that the U.K. government had, in fact, already decided to approve the use of land for transgenic crops — despite its £500,000 investment in public opinion on the subject, which overwhelmingly did not support that action. Any crop ban, according to the leaked documents, would be "the easy way out" and "an irrational way to proceed" in light of the government's desire to back and encourage U.K. science.[10]

≡

With this in mind, let's return to the mathematicians' characterization of innumerates: that we lack perspective about facts, about things that can be measured. Are innumerates really the only people who can be accused of this flaw? John Adams' seat belt safety researchers don't see impact, literal or figurative, beyond the bounds of the cars they design. Thurman's Nobel-winning economists doomed Long Term Capital Management by building their models on faulty assumptions, which they then relied on to make decisions; any other context was irrelevant.

So couldn't it also be true that mathematicians and scientists — and, significantly, the governments who rely on their counsel — can be just as limited, and even sometimes blinded, by their own beliefs, experiences and perspectives, as are we innumerates? Couldn't they be making risk decisions just like we do, based only on the information they've got, or what they choose to include in their analyses, ignoring any other possible context for the decisions they make?

The answer is *yes*, of course. And the approach isn't peculiar to those scientists who do risk assessment, either. In fact, it's the way scientific inquiry has been conducted since Francis Bacon fathered inductive reasoning. The practice is called many things, but the best metaphor for it may be "science under the lamppost." Based on the old joke about the drunk looking for his car keys under a streetlight ("Why are you looking there?" "Because that's where the light is"), science under the lamppost focuses on what it believes is the problem, and then sees or seeks the answer only within that circle of light. Anything outside the circle, no matter what its possible relationship might be to the outcome of the study, it ignores or dismisses.

While this may sound very much like Mendel's approach to his plant experiments, in practice it often turns out to be quite different. Mendel not only disclosed what he left out of his model as part of his report; he also disclosed that the interpretation of his results were contingent on where the light fell: i.e., the distinct traits of his plants. But it seems that many working scientists today have forgotten the critical factor of context in how they disclose and interpret their experiments to the rest of the world.

Today it's far more common for scientists not to reveal the thinking that went into designing their experiments. As Robert

Kohler wrote in *The Lords of the Fly: Drosophila Genetics and the Experimental Life:* "Scientists give facts an appearance of natural inevitability by veiling the process of their construction."[11]

That is to say: statistics, probability estimates, and risk analyses can all be objective within a given context or set of assumptions; in Adams' seat belt story, for example, that context would be *inside the car.* But "objective within a context" means the final answer is, by definition, subjective. If you change the context or the assumptions that underlie your model and measure again, the estimates *and whatever judgments you've made based on them* almost certainly will change, too.

And technical experts wielding numbers, with judgments and values and beliefs that shape both their perceptions and their conclusions based on those perceptions, are the ones who choose that context.

When an economist says there is a 30 percent chance of a recession next year, for example, he has no hard data to back up this statement. He's expressing a level of confidence in his own judgment, based on his own personal set of values and assumptions. He's 70 percent sure there won't be a recession, and this, like most probability estimates, involves personal judgment. But these imprecise, subjective probabilities often come with precise numbers attached to them. So where does the economist get his 70 percent figure? According to Thurman, the statistician, he would have based them either on empirical data, on data plus his own judgment or entirely on judgment.

"How do you think the Department of Homeland Security revises its terror alerts?" Thurman asked. Certainly not with data.

For a risk analysis to be meaningful, then, the process that produced it has to be *transparent.* It has to provide those who use the analysis with an opportunity to vet the model for themselves, to challenge its assumptions. If we're to be affected by some new technology, we should be able to ask about all the subjectivities and biases that scientists have built into their risk models. How might their biases have influenced how they modeled the problem? How did they influence the outcome? Even with its reliance on hard-core data, "experimental science could still be looking at completely the wrong things," said Thurman. "Your results are only as good as your model. The best way to mitigate this problem is by being transparent."[12]

But that is not the way most scientists prepare risk assessments today. Rather than providing us with a window to their world, they hold up a kind of funhouse mirror, reflecting only their own expertise and none of their judgment. In fact, there is such a "through the looking glass" quality to this whole operation, such a double standard by which subjective representations of reality are judged (specifically that the "experts" are correct and we are not), it's no wonder scientists believe that ordinary people don't understand risk. In the context of uncertainty, the reality is that most of them don't understand it, either.

≡

"Everyone is hooked on large amounts of data," said Warner North, a consulting professor at Stanford University and one of the best-known and most respected probabilistic risk analysts in the field. "But data are a very blunt instrument. What you can learn from data alone, from a decision or a risk perspective, is very, very limited."[13]

Trained in the deeply quantitative traditions of physics and mathematics, North is not one you'd expect to hear delivering indictments of an over-reliance on data. He's one of the earliest scholars and practitioners of decision analysis. Decision analysis is a quantitative approach to making decisions that was developed in the 1960s, specifically in response to the unique and complex palette of risks and benefits posed by nuclear energy as it made its bumpy transition from bombs to electrical power.

North also has served on more than a dozen risk-related study committees for the National Academies, of which he is a member. He has prepared extensive risk analyses for the broadest imaginable range of clients and subjects — from what was then known as the U.S. Atomic Energy Commission, to the Mexican government, the Grocery Manufacturers Association, and scores of other private companies and government agencies. He has been there and done it all, from the ivory towers of academic holding-forth to the dirtied fingernails and rolled-up shirtsleeves of working on real problems with serious consequences.

North and many of his equally distinguished colleagues have long been dismayed by the legions of academic and industrial scientists, corporate executives, and government regulators who continue to cling to the mathematical approach to risk analysis — despite its irrelevance to certain types of problems and sustained efforts by North and his colleagues to pry them loose from the practice.

Of course, a blunt instrument is sometimes the right tool for the job. Data-driven probability continues to be the solid foundation for several industries and disciplines, said North, most notably insurance and finance. It is and will continue to be the crux of their businesses, and as long as they continue to collect and use real-world data, that's as it should be. But the approach isn't useful everywhere. When it's used in the wrong circumstances or asks the wrong questions, it can lead to false conclusions — and bad, sometimes dangerous, decisions.

Assessments of the complex risks and benefits of technology innovations are especially vulnerable to over-reliance on data from a single point of reference, a phenomenon that one risk expert, Baruch Fischhoff, calls "the strongest available science." In the context of genetic engineering, the strongest available science is molecular biology.

Fischhoff, a professor in the Department of Social & Decision Sciences at Carnegie Mellon University, has long been attracted to problems where there are opportunities for unusual applications of social science methods and results, especially when they can be applied to problems in the real world. A prolific researcher and author of international reputation, as well as a member of the Institute of Medicine (and, for the 2005 term, president of the Society for Risk Analysis), he has extensively studied how scientists do their work, and how their methods and values affect the way risk is characterized and assessed.

The key word in Fischhoff's coinage is "available." Risk analysts shouldn't assume that the available data are the most relevant to the risk. In fact, he said, any risk analysis that automatically defers to the strongest available science without looking for relevant data from other sources is in danger of building its model on a faulty premise, which then misinforms the analysis, which in turn can lead to a bad decision.

Of course, it's much easier to assess risk using data from only one scientific approach or perspective, especially one like molecular biology that has the whiff of the Holy Grail about it. There are many reasons for wanting to stick to one kind of science, not least of which is the fact that the scientific community has no common language or methodology. Individual disciplines, even within biology, often use the same words to mean different things and have incompatible ways of measuring and presenting their results. Experts from different fields can have a difficult time just talking to each other, let alone arriving at a common understanding, even if they study the same topic. Anyone trying to model a risk that includes many different fields of expertise ends up having to figure out an equivalent of scientific Esperanto, just so that everyone can agree on the basic terms of the problem.

Nevertheless, a credible risk analysis for complex and complicated problems requires the effort. Otherwise the dominant science will end up unduly dominating the risk assessment. And that, as Fischhoff said, "can neglect potentially critical, deep uncertainties in other areas."[14]

When I talked to Harvey Fineberg, the head of the Institute of Medicine, I asked him how it might square with working molecular biologists, this idea that their experimental results might not be the strongest possible foundation on which to build a risk analysis. He replied that once again, the answer comes back to the question you're asking. Is it a research question, or a risk question? "To be useful to a decision maker, the strongest available science has to be both strong in the scientific sense, and strongly pertinent to the risk question. Knowing the real choices that have to be made guides you back into deciding *what's the most important science you need to develop.*"[15]

Those are my italics; that statement is really the crux of the matter. Modern-day risk analysis, particularly for new scientific endeavors and technologies like genetic engineering, is far more complicated than simply taking measurements and calculating the probabilities that a certain product or process is either safe or worth the risk. How you model the problem — starting with what you want to know, and why — and what data you use to populate your model can change dramatically based on whether you're trying to answer a research question or you're assessing risk.

Even if you've got the data, which is rarely the case for technological innovations, math experts don't consider the data generated by measurement to be the ultimate way to determine credibility and reliability. Here are a few caveats offered by the mathematicians Derrick Niederman and David Boyum, authors of *What The Numbers Say*:

- There is always more than one way to measure something.
- Measurements are error-prone.
- Even when correct, measurements are still only an approximation for what you really want to know.
- Measurements change behavior.[16]

The first three should sound familiar, as they are basically the same warnings against "mathophilia" that Thurman and North have already sounded. But the implications of that last simple statement, "Measurements change behavior," are profound. Like hatchlings attach themselves to the first object they experience, humans can't avoid being "imprinted" with the worldviews to which we've been exposed. Once we've learned to think or behave a certain way, without consciously retraining ourselves (and often even if we do) we seem unable to sincerely or easily follow any other course of action than the one that has been imprinted. "What gets measured gets done" is the relevant slogan in the context of statistics and probability.

"What gets measured gets done" is also pivotal in the argument for recalibrating how we describe and measure risks. Those who've been imprinted with the worldview that calculating frequencies in data (and, preferably, in *their* data) is the only credible way to gauge risk will find it very difficult to believe otherwise, even when presented with unequivocal evidence to the contrary. But the measurements that are generally part of a traditional risk assessment have a far greater impact than simply providing data for probability calculations. *Measurements also fundamentally influence the actions that we'll take in response to data* — they'll change our behavior — whether those actions are appropriate or not.

For example, the metric most people use to assess obesity is body weight. So imprinted are we with the metric of weight as an indicator of body fat that people who exercise in order to slenderize often quit a successful exercise program because they are "only" dropping clothing sizes rather than pounds. They can't wrap their minds around the reality that the number on the scale may have nothing to do with their progress; that in fact, they are building muscle, and muscle weighs more than fat.[17]

Other, more perilous behaviors can be triggered by the "measurements" provided by medical tests. An article published in early 2004 in the *New England Journal of Medicine* gave lie, for example, to a decade-long belief about a prostate cancer test. In this case, the test in question marked the line between what was perceived as normal versus abnormal body function.

Doctors have been operating (literally) on the belief that the standard line between normal and abnormal on the common blood test used to screen for prostate cancer was 4 nanograms of prostate specific antigen, or PSA, per milliliter. But results of the study in *NEJM* showed that no matter how low his PSA level, a man could still have prostate cancer. Until then, when a patient's test came back with a PSA level below 4, the doctor would decline to pick up the scalpel. How did "4" become the standard? No one, not even leading urologists, is exactly sure. One professor said it's because "doctors like whole numbers."

The bottom line, according to one of the doctors interviewed, is that they "just don't know what [the PSA] means."[18] So guess what else they don't know? How about how many surgeries were performed unnecessarily, based on that arbitrary "4"? Or how many people may have died because their doctors believed a sub-4 PSA didn't qualify them as a candidate for surgery? This phenomenon is not unique to the medical profession, although that's one place where its consequences can translate into tragedy.

From a risk perspective, imprinting usually takes the guise of this type of unexamined demand for a number, any number. It can be found almost everywhere you look. But for our purposes, it begs a very important question: *How do you know what to measure — or how do you even decide if there* is *anything to measure — when you're looking for the risks in something completely new?*

And just as importantly, what are the consequences if risk assessments use irrelevant or ambiguous data — in other words, if the analysts incorrectly assume that what they are measuring is what's important — to forecast consequences? Today's answers to these questions will not smooth your furrowing brow. Because if what gets measured is what matters, then its shadow is also true:

What doesn't get measured doesn't matter.

What might not officially "matter" in the risk assessments for genetic engineering? What are the models — the available science — that decision makers are using to develop their policies and regulations for the risks of transgenics? There's really only one place to start that conversation — with the very first officially approved food produced by recombinant DNA technology: the Flavr Savr tomato.

CHAPTER 6.

POLITICS, SCIENCE AND 'SUBSTANTIAL EQUIVALENCE'

"On May 18, 1994, the [U.S.] Food and Drug Administration announced that the Flavr Savr, a new tomato developed through biotechnology, is as safe as tomatoes bred by conventional means,"[1] trumpeted the FDA's press release, headlined "The Flavr Savr Arrives."

The Flavr Savr was a clever response to a problem that had long plagued commercial tomato growers. As the fruits ripened they also became soft, which made them difficult both to harvest and to ship so that they could arrive at their destination intact. Plant geneticists had discovered that "ripening" and "softening" are controlled by different genes. This led to the development of a new variety, produced by a California company, Calgene, Inc. The new tomato contained an extra copy of the same tomato gene that expressed the softening enzyme, but its position in the genome was reversed. This caused the Flavr Savr to produce less of the softening enzyme so that it could ripen normally, but stay firm enough to hold together for harvesting and processing.[2]

The new tomato marked the first time the FDA had evaluated a whole food produced by genetic engineering using its "Food Derived From New Plant Varieties" policy. The policy is better known as the doctrine of "substantial equivalence," which was codified into regulatory law in 1992. As its name reflects, the FDA policy treated transgenic food crops as just another type of domesticated plant.

Under the tenets of substantial equivalence, new crop varieties produced by using recombinant DNA techniques were considered to be essentially the same as the conventional varieties that are already "generally recognized as safe" (known by the acronym GRAS). In other words, any food or food product already considered safe that is produced or manufactured using transgenic corn (taco shells, high fructose corn syrup, corn starch and so on) or soybeans (soy milk, tofu, protein bars, soy lecithin and so on) would automatically be in the same category as a product using non-transgenic counterparts.

The FDA policy was explicit that the composition of the product itself was the important thing, not the method that produced it.[3] "[The FDA] is not aware of any information showing that foods derived by these new methods differ from other foods in any meaningful or uniform way, or that, as a class, foods developed by the new techniques present any different or greater safety concern than foods developed by traditional plant breeding," read the FDA's policy as filed in the Federal Register in 1992.[4] For this reason, the agency stated, the method of development of a new plant variety was not "material information" to risk.

Two years later, the FDA published a clarification about its definition of the term "ingredient" in the context of transgenics, again in the Federal Register (italics are mine):

> FDA considers an 'ingredient' to be a substance used to fabricate (i.e., manufacture or produce) a food. FDA does not consider those substances that are inherent components of food to be ingredients that must be disclosed in the food label. *A genetic substance introduced into a plant by breeding becomes an inherent part of the plant as well as of all foods derived from the plant ... regardless of the method used to develop the new plant variety. ...*[5]

In other words, no matter what the source of the new "genetic substance" — whether the transgene originated in an animal, a plant, or from some variety of microbe — it was of no material interest to the FDA's safety considerations. Consistent with its substantial equivalence policy, the FDA required no new safety tests, including feeding tests, to be performed on transgenic food. The only time a transgenic food needs to be labeled is if it contains an allergen people wouldn't ordinarily expect from the name of the food. As with all foods regulated by the FDA, it's up to the food producer to ensure the product is safe for consumption. In the U.S., neither the FDA nor any of the other agencies that regulate transgenics conduct independent allergy testing.

As the scientific basis for this position, the FDA used standard chemical tests comparing the *nutritional composition* of the two food varieties. Chemical analysis for nutritional composition doesn't typically reveal any substantial differences between, say, a conventional soybean or papaya or kernel of corn and its transgenic counterpart. For example, in 1996 Monsanto scientists published a detailed nutritional comparison of conventional soybeans versus its Roundup Ready transgenic variety. The study compared the two varieties' nutrient content, such as protein, fat, carbohydrates and fiber. It compared their anti-nutrients as well, which are the substances in food that can slow down the metabolism and absorption of nutrients. The Monsanto study even analyzed and compared the amounts of the amino acids that made up the proteins in each variety.

At the end of the study, the researchers found that the Roundup Ready soybeans showed no significant differences from traditionally bred soybeans, either in composition or amounts of these various substances.[6]

But as the statistics and math experts in the previous chapter cautioned, what if nutritional composition, like the PSA level for prostate cancer, isn't relevant to the hazard that transgenic foods present? What if the genetic engineering process actually is material to the safety of transgenic food? The story of where and how the FDA got the authority to make its claim gives new meaning to the term "political science." It also is distressingly revelatory of the process by which regulators decide to officially assess risks today.

≡

Once the early risk issues had been "settled" at Asilomar and the patent for recombinant DNA was awarded in 1980, several biotech companies were quickly founded. (The first, Genentech, had already been launched by Herbert Boyer in 1976.)[7] It was not lost on those at the forefront of the science that in addition to whatever revolutionary new discoveries might result from recombinant DNA, the newfound ability to create wholly new "varieties" of hybrid organisms could also launch a revolutionary new industry. Neither was this fact lost on the Reagan administration in the mid-1980s, for whom the budding biotech industry had taken on heightened significance.

The U.S. had long been the world leader in science and technology. Its most recent big success at the time, the semiconductor, had created an enormous new international market for computer technology and a budding "information revolution." But in the mid-1980s the Japanese had already beaten U.S. companies in one important area, memory chips, and it was feared that they were on their way to capturing the entire market.[8] The U.S. chip industry's woes were coincident with the early years of the biotech industry, and when it looked as though the U.S. might be losing ground in the world economy, biotechnology became the cause célèbre for continuing U.S. industrial prestige.

Reagan's policies always pushed to deregulate American industries, but Monsanto asked the Reagan White House to take a different tack: it asked for its new transgenic food crops to be regulated, and it wanted the Reagan White House to proactively champion the idea.[9] As a result, three important pronouncements came from Washington:

First, in 1986 the White House Office of Science and Technology Policy (OSTP) published the "Coordinated Framework for Regulation of Biotechnology." The "Coordinated Framework" proposed that the products of biotechnology, created by the process of genetic engineering, be regulated — but only under existing laws.

"By product, not by process" was the catchphrase that OSTP used, meaning that the process used to create the products (i.e., genetic engineering) was not to be considered unique or important from a risk or regulatory perspective. Only the characteristics and novel features of the transgenic product itself were to be subject to safety regulations. The express purpose of the "Coordinated Framework" was to provide "a measure of regulatory certainty for industry," according to the White House announcement, and to put the nascent industry "on the fast track for commercialization and international competition."[10]

Then, in 1989, what's now the National Academies assembled a committee to study the safety of field-testing genetically modified organisms. Members of the committee included several scientists either affiliated with or employed by biotech companies. Not surprisingly, the committee agreed with OSTP's pronouncement that biotech regulations should focus on the "product, not process," and it published a much-cited report to that effect.[11]

Finally, in 1991, the Council on Competitiveness, chaired by Vice President Dan Quayle and explicitly aimed at reducing government regulation, joined the chorus for a "product not process" approach to biotech regulation. When he introduced the FDA's final policy for the White House in 1992, Quayle referred to the substantial equivalence doctrine as "regulatory relief" for an industry that as yet effectively *had* no regulation — and, as yet, no products.

But if the decision to focus regulatory attention only on transgenic products and not the genetic engineering processes that created them was another "bold experiment in industry-government cooperation," there certainly wasn't a united front within the FDA on the terms of engagement. Internal documents written by FDA scientists who contested the policy in advance of its approval show that the FDA's statement in the Federal Register — that it was not aware of safety issues regarding transgenic food — was simply not true. Many senior FDA officials and scientists who were consulted about the policy, ranging from compliance officers to the director of the Center for Veterinary Medicine, voiced serious concerns over the safety of both genetic engineering and its products. They particularly protested the conclusiveness of the science that the policy document purported to represent.[12]

These internal concerns came to light when a consumer group sued the FDA in 1998 to try to compel the agency to perform mandatory safety testing on transgenic foods. Documents gathered from the pre-trial discovery process and posted on the Internet revealed for the first time the concerns of the FDA's own scientists about both the doctrine of substantial equivalence and the presumption that transgenics should be classified under the category of "generally recognized as safe."[13]

The most eloquent and revelatory of the documents were memos sent from Linda Kahl, then the FDA's compliance coordinator, to James Maryanski, its biotechnology coordinator, directly commenting on the draft Federal Register document. Kahl's sensible approach perfectly captured the problem with the entire endeavor — a problem that continues to be epidemic with respect to how risks are assessed for science and technology regulations.

She began by questioning the objective of the policy statement. "I believe there are at least two situations relative to this document in which it is trying to fit a square peg into a round hole," Kahl wrote. "The first ... is that the document is trying to force an ultimate conclusion that there is no difference between foods modified by genetic engineering and foods modified by traditional breeding practices. This is because of the mandate to regulate the product, not the process."

Kahl didn't say whose mandate she was specifically referring to, although based on the OSTP's public recommendation to that effect in 1986 — six years before the FDA policy was announced — it seems fairly obvious that it came from the White House. She continued, "The processes of genetic engineering and traditional breeding are different, and according to the technical experts in the agency, they lead to different risks. There is no data that addresses the relative magnitude of the risks — for all we know, the risks may be lower for genetically engineered foods than for foods produced by traditional breeding. But the acknowledgment that the risks are different is lost in the attempt to hold to the doctrine that the product and not the process is regulated."

The other "square peg" was "the approach of at least part of the document is to use a scientific analysis ... to develop the policy statement," Kahl wrote. "In the first place, are we asking the scientific experts to generate the basis for this policy statement in the absence of any data? It's no wonder that there are so many different opinions — *it is an exercise in hypotheses forced on individuals whose jobs and training ordinarily deal with facts.*" (My italics.)

In the second place, she wrote, "I don't think that the scientific analysis as presented is complete. ... I believe that in most cases the word 'risk' is avoided. This is probably at least partly due to the fact that there is no data that could quantify risk. ...

"I wonder if part of the problems associated with this approach — using scientific issues to set the stage for the policy statement — are due to the fact that the scope of technical experts assigned to the project did not include any whose usual job is risk analysis. This does not eliminate the problem with a lack of data, but if the molecular biology, chemistry, and toxicology experts are being forced to deal with hypotheses rather than data, why not the risk analysis experts?"[14]

Good question. Kahl's description of the agency's approach to the problem dovetails perfectly with Harvey Fineberg's earlier comments about why scientists are so bad at risk analysis. It seems especially relevant since risk analysts are trained to make decisions in situations precisely like these: in which scientific data is sparse and uncertainty is abundant. Anyone who wants to do risk assessments in the realm of science and technology needs to learn the tenets of a very different discipline, one that is steeped in the methods of decision making under conditions of uncertainty. Even then, the lack of data makes the process extraordinarily difficult.

That is not to say that the FDA had no data when it came to approving the Flavr Savr under its new substantial equivalence policy. The company that produced the transgenic tomato, Calgene, behaved with admirable responsibility and actually submitted the results of feeding tests for Flavr Savr, even though it wasn't required to do so under the directives of the FDA's substantial equivalence policy.

One of Calgene's three rat feeding studies was not reassuring: stomach lesions were observed in four out of 20 female rats fed the Flavr Savr tomatoes. Yet in its report, Calgene deemed the lesions to be incidental, which in medical parlance means "of no clinical significance."[15]

In response to the Calgene report, an FDA staff pathologist had said in an internal memo, "There is considerable disparity in the reported findings of gastric erosions or necrosis lesions from the three studies provided by *Calgene Inc.* This disparity has not been adequately addressed or explained by the sponsor or the laboratory where the study was conducted ... The criteria for qualifying a lesion as incidental were not provided in the Sponsor's report."[16]

Nevertheless, the official word from the FDA's Center for Food Safety and Applied Nutrition concluded that the new variety "had not been significantly altered in regard to safety," and it was approved with no further tests required.[17]

Where is the Flavr Savr today? Calgene stopped producing the recombinant variety in 1997. Although industry mythology likes to blame this on anti-biotech activism (of which there was plenty), the reality was much more mundane: Calgene's scientists had picked the wrong type of tomato to genetically engineer, and as a result, crop yields and quality were far below expectations. Only about 10 to 20 percent of the tomatoes were good enough to merit the premium brand label that Calgene put on the bioengineered variety.

And ironically, the tomato's raison d'être of "firm yet ripe" didn't prove to be such a good idea after all. While Flavr Savr technology did delay the tomatoes' softening, they were still more delicate than the traditional early-harvest, hard green tomatoes. Many were ruined before they reached Calgene's distribution center. Altogether Flavr Savr was costing the company a fortune that it was not recouping in sales.[18] Meanwhile, between 1995 and 1997, Calgene was bought up by Monsanto, bit by bit, and scientists began moving the genetic ripening trait into other premium tomato varieties.[19]

I'll admit to a perverse admiration for the substantial equivalence doctrine, a "science"-based policy declaring that no policy is required. It was genius, really: if the measurement you use to evaluate transgenic food "proves" that it's legally the same and as safe as what everyone already eats, why would a risk analysis *ever* be required? Why would you label those foods, or monitor who or what eats them? You wouldn't. In fact, you could even actively lobby against labeling transgenic food, calling it scare-mongering and anti-competitive. Without labels or monitoring, if transgenic food does turn out to be a hazard, there's no obvious way to track the problem back to its source. Thus with no paper trail of liability for either the agency or the developer, no one has to "own" the risk. Most brilliantly, everyone *except* the agency and the developer will have to pay the price: consumers, food producers and distributors, the health care system, the economy, the environment.

In the perfect coda to this story, the legal challenge that brought the FDA's internal safety debates to light was dismissed by the U.S. District Court judge assigned to the case. She said there was no data "proving" that transgenic foods are different from traditional crops — "no data" being precisely the reason for caution that Linda Kahl presented in her critique of the substantial equivalence doctrine. But the judge came to a different conclusion about that fact than did the FDA scientists. In the absence of data, she claimed, the administrators of the

FDA are legally entitled to establish whatever policies they'd like — whether or not their own scientists agree with them.

The FDA was not alone in its embrace of the "product not process" approach to transgenic policy and regulation. The USDA, for example, had already approved Flavr Savr to be grown in open farm fields long before Calgene needed the FDA's approval to sell the tomato. Upcoming chapters will look at methods by which both the USDA and the EPA in the U.S. have used this approach for their own transgenic regulations. But what might the risk landscape look like if regulators *were* to include the process of genetic engineering, and not just the product, in their risk considerations?

≡

The traditional plant breeding methods to which genetic engineering is compared use cross-fertilization and cross-pollination in much the way Mendel did, to impel a desired trait to show up in the next generation of the plant. The problem with this process, however, is that it's slow: it ends up mixing thousands of genes together, and breeders can spend years selecting and rebreeding generation after generation to keep the genes they want and push out the genes they don't. According to an article in the FDA's consumer magazine, published in 2003, the benefit of using genetic engineering instead of these traditional methods is that it is "more precise and predictable ... a single gene may be added" to a plant to give it a single specific characteristic without transferring the undesirable traits.[20]

This sounds like a terrific idea, and it's how genetic engineering is usually described and understood: as a simple process that involves splicing one foreign gene, representing one desired trait, into the genome of a new host. The transgenic protein in Roundup Ready soy, for example, is an enzyme that originated as a sequence of DNA found in *Salmonella* bacteria. Scientists discovered that this particular bacterium's cell walls could survive the application of glyphosate weed-killer in much the same way that some bacteria can survive the application of antibiotics. As mentioned earlier, the second-most popular transgene, *Bt*, came from a soil bacterium that produces a toxic protein that kills specific kinds of insects when consumed.

But the FDA's "single gene" description is simply wrong. Splicing DNA is not like splicing an electrical cord, or cutting a single word out of one document and pasting it into another one. Genetic engineering requires an entire battery of chemical tools and the assemblage of an entire construct of DNA sequences to impel that single gene to lodge and operate in the genome of a wholly new organism.

This transgenic "cassette," as the construct is called, contains not just the single target gene. It also contains at least one and usually more DNA sequences from viruses or bacteria, chosen because of their powerful and well-demonstrated ability to do two things that are required by genetic engineering: one, to cross species lines, in order to infect the cells of other organisms, and two, to commandeer the cell's genetic replication mechanism, in order to reproduce their own DNA's traits.

So how does genetic engineering use these powerful infective agents?

Inside living cells, the physical "characters" and traits that are being swapped from one organism to another generally take the form of protein, and can often be linked to individual sequences of DNA. Virtually everything in a cell that's not water is protein. Enzymes, for example, which trigger most of the chemical reactions in the body, are proteins. So are hormones and antibodies, and structural substances like collagen.

Proteins are also central to the discovery of the "standard genetic code." The code is a set of rules that governs how the information contained in sequences of DNA and RNA is translated into proteins. It applies to virtually all living cells. The discovery of a standard code for proteins is what laid the groundwork for the invention of recombinant DNA. If the same sequence of DNA expresses the same protein in any living cell, no matter what the organism, then it stands to reason that transplanting that sequence should be able to produce its coded trait just as well in the genome of a new host.

Very crudely stated, that's the goal of the genetic engineering process: to find the DNA that expresses a desired trait, cut that specific sequence of DNA from one organism's genome, then paste the sequence into the genome of a different organism so that the new host produces the new trait.

But many different sequences of DNA are involved in bringing that process to fruition. Biotech companies publish at least a partial list of the contents of their products' transgenic cassettes as part of what's generally a voluntary disclosure process for regulators, including the FDA. The list for Monsanto's Roundup Ready cassette,[21] for example, starts out with the gene for glyphosate resistance. Once it's chemically snipped from the *Salmonella* genome using a "cutter" enzyme (an enzyme now synthetically produced in the lab by cloning), getting the glyphosate resistance gene into a plant cell requires building a kind of "DNA shuttle" to penetrate the nucleus of the host cell and drive the target gene directly into its new genome.

DNA shuttles are generally made from pieces of bacterial DNA called plasmids. Because plasmids reproduce outside the cell, they can shuttle bacterial DNA — typically genes that cause disease — into the cells of wholly different organisms. They're also the vector that transfers antibiotic resistance genes between species of bacteria. In the case of Roundup Ready soy, the plasmid is harvested from *E. coli* bacteria. Another popular plasmid in genetic engineering comes from *Agrobacterium tumefasciens*, called *Ti*, which in its infective state causes tumors in plants.

To use a plasmid for genetic engineering, scientists first chemically "defang" the plasmid, removing its actively infective or dangerous genes. In order to replace these genes with the transgenes they want to shuttle, they open up a gap in the plasmid and paste the new gene into place using another cell enzyme that binds DNA pieces together. Several other genes are chemically "pasted" into the transgenic cassette before it gets launched into its new host.

Another critical component of the cassette is what's called a "promoter" sequence of DNA. The process by which a gene's DNA sequence is converted into protein is something like reading a sentence from beginning to end. The sentence begins with a promoter sequence that enables the target gene to start "expressing" or creating its protein in new surroundings. Virtually all commercial transgenic plants use the cauliflower mosaic virus promoter, because it acts as a kind of supercharger for the target gene, getting it to express more of the protein than it would normally (which you'd expect from a virus trying to boost the volume of its infectious bits into a new cell). Another bit of DNA from an *Agrobacterium* is used as a terminator, which stops protein expression.

There are generally other bits of DNA included in transgenic cassettes that are designed to perform various other functions, like impelling the target protein to express in certain parts of the plant (or animal) and not in others. In Roundup Ready, this bit of genetic material comes from a petunia, for example. Until recently, virtually all commercial transgenic cassettes have also included a sequence of antibiotic-resistant DNA from the *Streptococcus* bacterium.

Once the cassette is constructed, it's cultured in the lab to make millions of copies. The bacteria, loaded with transgenic plasmids, are cut loose on targeted plant cells. The plasmids shuttle from the bacteria over to the plant cells in order to "infect" them with the transgene. (Alternatively, the cassettes can be blasted across the cell wall by means of a "gene gun," a device much like a pellet gun.) If the cassettes make it into the host cells' nuclei and burrow into their chromosomes, they are home free. The host no longer sees the new sequences as invaders, and makes room in its genome for some of them to "recombine" with its existing sequences.

The final step is to find out whether the procedure worked. To find out which cells "took up" the transgenic bacteria, genetic engineers douse them with antibiotics. Because antibiotic resistance genes are part of the transgenic cassette, any cell that survives the dousing has made room for at least one copy of the cassette in its genome. Plant cells that have taken up the cassette are then cultured into mature plants, which bear transgenic seed. A successful outcome means that every cell of the new transgenic organism contains *at least one copy* of this complete construct. And therefore all the plant's cells are expected to produce all of these new proteins continually, as part of normal operations.

≡

So much for the "single gene," cut-and-paste version of the process of genetic engineering. As for the FDA's claim about the precision and predictability of the process, that's pretty far off base as well. Besides the fact that it takes a village of DNA to infiltrate a host cell, it often takes thousands of tries to get a DNA sequence to fix in the proper part of the genome once it gets there.

In fact, botched transgenic experiments, where something in the process has visibly gone awry and the appearance or behavior of the organism has been unpredictably compromised, are more the rule than the exception. This is accepted as an inescapable artifact of the process. First-generation transgenic plants that look deformed, odd or otherwise not like they're supposed to are constantly screened out by genetic engineers and are usually destroyed. Only normal-looking transgenic plants make it to the next stage of product development.[22]

But normal-looking isn't normal. Even if a plant isn't visibly deformed, the process of transgenesis that I just described very often leaves not just one, but several copies of the transgenic cassette, or fragments of it, in various places in the plant's genome. To date there has been no effort to track or evaluate the possible health or environmental effects that these more subtle mutations might trigger in transgenic organisms. These subtle mutations — the effects of which don't necessarily rise to the level of a full-blown allergen or toxin, or that may take many years to manifest themselves — are completely unaddressed by the present methods of risk assessment.

Perhaps these subtle changes produce proteins that give you a low-grade stomach ache or change the thickness of the walls of your intestines. Or perhaps they're sending out a new chemical that gives you a headache or makes your tongue itch. Maybe some bit of transgenic DNA or protein is interacting with a chemical or microbe in water or soil in a way that slightly changes its acidity or alkalinity, and this in turn affects how well crops grow or creates a welcoming environment for a new kind of microbe in the water supply.

There is nothing about the substantial equivalence policy, nothing about focusing only on the nutritional equivalence of transgenic food products, that could or would provide a way for decision makers to anticipate or take notice of these types of effects. For a vivid example, let's look more closely at the potential for allergic reactions to transgenes, possibly the longest-standing concern about transgenic foods since they were first introduced.

Genetic engineering introduces proteins not normally found in conventional varieties of plants and animals. Thus transgenic varieties could conceivably trigger food allergies in one of two ways: either by expressing proteins from foods that are already known to trigger allergic reactions, like eggs or peanuts or shellfish, or by introducing proteins completely new to the food supply that cause unprecedented reactions.

The first issue seems easy enough to control, based on at least one prior study. In 1996, a research team wanted to boost the nutritional value of soybeans by inserting a Brazil nut protein into the soybean genome. The blood serum of people with Brazil nut allergies showed an allergic reaction to the soybean extract, so the product was never brought to market.[23]

But it's the issue of novel proteins no one is testing for that's of much greater concern from the standpoint of food safety. This was one of the fears in 2003 when StarLink, a transgenic corn approved for animal feed, found its way into taco shells, corn chips and other products that people eat. StarLink had been approved for use in animal feed by the EPA, which regulates transgenic pesticides, but not in products intended for people, because testing revealed that the protein was resistant to being broken down by digestive juices in humans — a key feature of many allergens. More than 200 people reported having allergic reactions to packaged foods containing the corn, which moved the problem from the EPA to the FDA's jurisdiction, although government scientists said they didn't find a clear link between the food and the reactions.

Opponents of transgenics criticized the government's methodology for the study, claiming it ignored "suggestive evidence of allergenicity"[24]; one vociferously pro-biotech organization accused these "nanny groups," as it called them, of "desperately looking to poke holes" in the FDA's testing methods "so they can continue their scare tactics."[25] Either way, it may be years before all StarLink is taken out of the human food supply.[26]

More recently, in November 2005, a decade-long project in Western Australia to develop transgenic peas with built-in pest resistance was abandoned after tests showed that the pesticide protein caused an allergic reaction in mice, resulting in lung damage.[27] What's scientifically noteworthy about the case is that when the protein was expressed in its home genome — a variety of common bean — it did not cause allergic reactions in either mice or people. But when it was recombined with the genome of the pea, its structure changed subtly and the unexpected immune effect occurred.

What if that pea had been developed and marketed in the U.S.? Under the substantial equivalence doctrine in the U.S., which is considered the baseline regulation for transgenic foods in many countries around the world, it would have been considered a food that was "generally recognized as safe." The company developing it would have been under no obligation to test it as an allergen.

Recognizing this rather significant oversight, the Western Australian government announced that it will fund independent safety studies on transgenic food crops, including animal feeding trials. "There has been a concern for a long time that when a gene is taken out of one organism and put into another, the protein expressed by that gene may be different," said Kim Chance, the agriculture minister of Western Australia, in a statement issued shortly after the news broke on the transgenic pea.[28]

The other ticking bomb is our lack of consideration for hazards that may have long gestation periods. The carcinogen in DES, for example, didn't manifest until its victims became sexually mature, and so the mutations it caused have been passed down to children in every generation since its inception. A more recent example — and one more relevant to serious food safety concerns — is variant Creutzfeld-Jacob Disease, the human variant of the prion disease TSE. Unlike the gestation period for many food-borne diseases, the 100 percent fatal vCJD can gestate for anywhere from three to 50 years. So far, the USDA has actually prohibited beef producers from testing all their cattle at slaughter, for fear of sending "the wrong message" about the safety of U.S. beef.[29]

≡

Nearly 15 years after the Flavr Savr was approved by the FDA, ongoing concerns among consumers and scientists finally prompted the FDA, USDA and EPA in 2004 to revisit the issue of transgenic food safety. The agencies commissioned a new National Academies report, the first since the original 1989 study on field-testing transgenic crops, on the safety and unintended health effects of transgenic food.[30] Of course they didn't call it "transgenic food" in the study title. They called it "genetically engineered food," so they could include all methods of genetic modification in their consideration and not be seen as singling out transgenics.

But no matter what its name and no matter what public-relations spin the authors or the biotech industry put on it, a close reading of the study's results drives a stake into the heart of the substantial equivalence argument for regulating only the products and not the process of genetic engineering. Buried under several layers of political double-, triple- and quadruple-speak, the study committee concluded that any breeding technique that alters a plant or animal — using recombinant DNA or other methods — "has the potential to create unintended changes in the quality or amounts of food components that could harm health," according to the committee chair, Bettie Sue Masters, a distinguished chemistry professor from the University of Texas Health Science Center in San Antonio.[31]

The committee made several sensible recommendations for which we should all be grateful, if they ever get implemented. They would dramatically improve the way food safety and risk are assessed.

For example, the committee recommended that all the agencies responsible for various aspects of food safety in the U.S. conduct an appropriate safety assessment of these foods *before* they allow the products to go to market. That one is 15 years too late, but is nonetheless gladly noted.

It also suggested that the agencies stop using outmoded metrics that ignore the process used to produce food. They should step into the 21st century and take advantage of existing scientific techniques for analyzing and profiling the composition of food. Along the same lines, it also encouraged the agencies to support the development of better tools to "detect and assess the biological consequences of unintended changes" in modified food.

Most notably, and at long last, it recommended that the agencies figure out a way to monitor and track adverse health effects from transgenic foods once they've already entered the market.

Overall, the message to the agencies that regulate transgenic food was "step up to the plate." Committee members declared that the agencies need to make a significant effort to detect the changes not only in transgenic food, but in the human and animal populations that have been consuming them for the past 15 years.[32] And until they do, the unintended effects of the government's long-standing and willful shortsightedness about transgenics will continue to accrue.

CHAPTER 7.
SILENT GENES

No matter how wise or authoritative they might be, recommendations by even the most august scientific advisors are not policies. At the time of this writing, none of the recommendations of the Academies' 2004 food safety committee had been adopted by any of the agencies that commissioned the study. Until new analytical methods and detection tools are developed and codified into law, we'll still be at a near-total disadvantage if we want to determine the risks that may be accruing in transgenic foods, or any other kind of transgenic organism in wide circulation. Given historical precedent, it's extremely unlikely that changes are imminent. As in so many other circumstances where regulatory law meets industry, the lip service and heel-dragging can go on for years.

In the meantime and as a result, the transgenic free-for-all continues around the world. Transgenic organisms are in wide circulation and are firmly established in the ecosystem. Any damage these organisms may be causing is invisible by today's risk assessment practices and irreversible by today's technologies. Such promiscuous, casually regulated use of powerful biotechnologies should beg the question: How might someone exploit the laxity of transgenic regulations as a weapon for bioterrorism? Could genetic engineering be used as a weapon?

I asked Jack Heinemann, the New Zealand biosafety researcher who talked about the threshold effect in Chapter 3, for an example of how such a scenario might involve the food supply, where transgenics are most common and plentiful. Because he regularly re-evaluates official risk analyses that are submitted to government regulators, Heinemann has isolated a set of special risks that transgenics present for food safety in particular. These risks hold true, he explained, whether the problems were created inadvertently or purposely engineered.[1]

By way of example, he recalled an event that made international news in 2003, when a pizza-topping company in Japan performed a routine genetic test on a shipment of New Zealand-grown corn it was using for its product. The test detected transgenes from the *Bt11* strain of transgenic corn sold by the Swiss biotech conglomerate Syngenta AG.

At the time, all commercial transgenics were illegal both in Japan and in New Zealand. The New Zealand company that bought, planted and harvested the corn thought it was using traditional seed, not a transgenic variety. Believing it was in compliance, it shipped the corn to Japan. The contaminated corn was returned to New Zealand and once it arrived, more sensitive tests on the seeds confirmed the Japanese company's findings.[2]

How Japan found the transgenes and New Zealand didn't is an important factor in this cautionary tale.

Japan is notoriously security-conscious about its food. It was one of the first countries to cut off U.S. beef imports after mad-cow disease was detected, for example, and maintained the ban for two years. At the time this went to print, the Japanese government had instated a new ban, on the sale of food products containing U.S. long-grain rice, after the discovery that a transgenic strain had contaminated commercial supplies.[3] As a result of this scrupulous attention, Japan's scientists have become very good at detecting what Heinemann called "little signatures" of genetic modifications and variations in foods. Japanese researchers were first to discover that eight different variants exist for the cauliflower mosaic virus promoter that's present in most commercial transgenes.[4] This fact is noteworthy in the context of food security because the mosaic virus sequence is the most common component of any commercial transgenic cassette. As a result, it's the easiest and usually the first thing that regulators look for when testing for transgenes. The terminator from a soil bacterium, called *nos*, is probably the second most common.

According to Heinemann, the standard test in most countries for transgenes in food looks specifically for one of these two sequences, starting with the virus promoter. If a sample tests positive, then the organism is assumed to be transgenic. "If it doesn't test positive for either the mosaic virus or *nos*, the sample is assumed negative for GMOs and pushed through to approval," he said.[5]

But the test that looks for these sequences is exquisitely sensitive, and as a result is very prone to error. Each test must be set up with absolute precision in order to produce accurate results. To get a solid "positive" for the presence of any variant of the mosaic promoter, there has to be a *perfect* match between the promoter and a custom-made chemical called a primer that triggers the detection test to respond. If the primer is off *by even a single molecule*, the test will either yield a weak result — which is often dismissed as an artifact or error in the testing process — or show nothing at all.

"But the Japanese don't assume that a weak signal is an artifact," said Heinemann. "They cut out the piece that produces the weak signal and test it again with a new, customized primer. They do that over and over until they can identify and describe even very tiny sequences of DNA."

In this case, their scrupulous attention yielded a positive test for contaminated corn — and a new awareness of the vulnerability of New Zealand's corn market.

But even custom primers aren't going to discover transgenes that someone doesn't *want* detected, which makes the larger implications for food safety even more disturbing.

To slip a transgene past a standard detection test in a country where transgenics are legal, "the smart thing to do would be to mix in my bad transgene with a known commercial one" that used the standard promoter and terminator, said Heinemann. As noted, once a government agency detects either one, they generally stop looking for anything else. Conversely, to sneak a "bad" transgene into a country where transgenic crops are illegal, "all you'd have to do is design it *without* using either of those two sequences, using a different promoter and terminator, and most of the tests done by regulators today will never detect it," said Heinemann.

There's another easily exploited weakness in the detection techniques used today. The computer programs that analyze detection data look for specific stretches of DNA that begin with a promoter like the mosaic virus and end with a terminator like *nos.* These stretches, called "open reading frames," usually indicate the presence of an active gene — that is, one that codes for a protein. "But if you know how the computer looks for things, you know what it doesn't look for, too," said Heinemann. "For example, until extremely recently, these computer programs wouldn't look inside one frame for another one. So someone could easily create a commercial look-alike transgene with a whole different transgene nested inside of it. Unless you look at the data and not just the report from the computer program, you won't find the one that's hidden. And the hidden gene could be a Trojan horse" designed with bad intent.

By any definition, this is bioterrorism. To possess and use the chemicals and materials required to assemble DNA molecules for this purpose are clearly illegal under the U.S. Patriot Act of 2001, which requires that whatever biological agents or delivery systems are concocted be "reasonably justified by ... peaceful purpose."[6] But the Patriot Act is certainly easy enough to get around, especially if you don't live in the U.S. No one has to deploy anything like anthrax or smallpox to wreak serious havoc — all they need is knowledge of the gene and the genome they're targeting and a few common chemicals. It's certainly not any more difficult to create these transgenes for the wrong reasons than for the right ones.

How might someone use this kind of knowledge? A criminal could use toxic transgenes as a weapon to kill a lot of people. But a more likely tactic, while lacking the drama of releasing sarin or anthrax spores on a subway, would be to target *economies* instead of people. "Economies are not generally the single-minded pursuit of the kinds of terrorists we usually think about," said Heinemann. "But monkey-wrenching each other's economies is very common among nation-states as a competitive 'technique,' and also between rivals in industries."

One clever way to do this might be to use transgenics for what Heinemann calls "remote control" of an industry. Consider, for example, the dairy industry. If a competitor wanted to use transgenics to play the industrial sabotage game, finding a way to decrease a big dairy company's milk production by even a small increment would have a significant effect on its ability to compete in the market.

"Let's say I figure out how to modify the food that a dairy company feeds its cows. This modification is designed to move from the food source into the cow's cells and reduce a certain protein responsible for milk production, so their productivity goes down just a little," said Heinemann. "Maybe I do that by modifying a soil bacterium with a gene that reduces their milk production and engineering it into their feed. Cows eat dirt, and lots of cows already eat feed that contains DNA from *Bt*, which is a soil bacterium, so no one would suspect anything was wrong if they did detect some bacterial DNA in a milk sample. But get this altered feed to them over a couple of seasons and the drop in productivity would change the market dynamic."

What's noteworthy about this scenario is that unlike a more traditional terrorism tactic, the intent is specifically *not* to be noticed or detected. Given the shortcomings of today's detection methods, as noted by the National Academies study in the last chapter, the offending transgene would almost certainly go undetected. And if it were detected, by that time some degree of harm would already have been done.

"Traditional terrorists will do the big dramatic thing," said Heinemann. "GMOs might be useful for that, but how they could be extremely useful is to let industry players modify the wealth potential of other players over really long periods of time. Because those people don't want to get caught. They just want to make more money."

≡

A mechanism called "RNA interference" provides a highly plausible explanation for how serious such threats could someday be. There have been plenty of good scientific reasons over the past 30 years to undertake a serious re-evaluation of the risks of genetic engineering, but the discovery of RNA interference in 1998 may be the most compelling. Also known as gene silencing, the mechanism of RNA interference is of such fundamental importance to understanding the biology of the cell that by 2006, it had already yielded the Nobel Prize for the two scientists who were the first to fully describe the process.

One of several potent "genome defense" mechanisms employed by cells, RNA interference kicks into action when a double-stranded RNA molecule is formed inside a cell. Recall that DNA is double stranded, but RNA generally isn't. One way that double-stranded RNA can arrive in a cell is when the cell is invaded by a virus trying to commandeer the cell's DNA in order to reproduce. (Viruses are often just a shell of protein around a bunch of double-stranded RNA molecules.) When a cell detects a double-stranded RNA molecule, it sends out enzymes to attack and dice up the invader. In what is the cellular equivalent of racial profiling, the cell then gathers up the scraps and uses their precise sequence as a template to search out and destroy every other RNA in the cell that matches the template.

This swift action on the part of the cell is brilliant at preventing infective or otherwise harmful bits of genetic material from replicating inside the cell. But this "interference" with the activities of RNA in the cell can also snuff out legitimate messenger RNAs that translate DNA into protein. Snuffing a messenger RNA stops the production of that protein almost as effectively as if its DNA had been physically snipped from the genome.

What's more fascinating is that RNA interference isn't just a cell defense mechanism. It's also a critical component in the development of an organism. Except for some sex cells, each cell of every living organism contains its entire genome, a complete sequence of DNA, capable of producing every one of the proteins it needs throughout its life, to sustain its life. But we don't need or want our stomachs to produce the proteins that, for example, make our brains or eyes function. RNA interference seems to be part of the control mechanism that regulates which genes are active in what kinds of cells. Although very little is known about the process, these regulatory networks also put some genes on hold while we're young, others when we're older, and so on.

Given its critical role both in genome defense and biological development, it's easy to see why the mechanism of RNA interference is one of the most important discoveries in cell biology. It has given rise to an entirely new area of study, with scientists striving to quickly perfect the ability to selectively silence the genes that produce "troublesome" proteins — like those that cause Huntington's Disease, for example, or infective prions in livestock.[7] And while the goal is laudable, the desire to productize such a fundamental discovery seems a bit premature. The science that can explain what activates or represses the silencing mechanism is virtually nonexistent, as is knowledge of how (or why) genes interact with each other, or the effect that silencing one gene might have on the rest of the organism.

But there are other things about the phenomenon of RNA interference that are worth noting here, beyond how little we know about its larger effects.

First, as with the classic, "single gene" description we've been getting for the past few decades about genetic engineering, gene silencing may only *appear* to be highly specific. Taking a narrow enough view, a genetic engineer can cook up precisely the right bit of RNA in the lab to inactivate just about any known gene or trait. The big caveat, as noted, is that there's little known about what inactivating one gene might inactivate, or even trigger, elsewhere. In fact, it's worth noting here that for the past two decades, we've been reading about "the gene" responsible for producing various diseases in individuals. We've heard about links to diabetes, Alzheimer's, obesity, schizophrenia, depression, and many others. But in early 2005, a *Wall Street Journal* columnist published a report that as many as 95 percent of these gene-disease links in fact fall into the category of the dreaded "false positive" result. That is, researchers wrongly made a cause-effect connection that doesn't exist between a gene and a disease.[1]

One comprehensive analysis, according to the *Journal*, counted more than 600 associations reported between a DNA variant and a disease. Only six had been consistently replicated. And because science journals don't like to print negative results, most ordinary people — and who knows, maybe even most scientists — still believe that single genes are strongly correlated with single functions.

Also, the research so far seems to show that these short pieces of RNA can move easily to and from any organ or system within an organism — and *between* organisms and species as well. Studying how mobile and how sophisticated these small double-stranded RNA molecules are has led researchers to discover that they can be ingested, survive the digestive process and still maintain their ability to silence a chosen target. Such hardiness means gene silencers will be easy to "package" and administer as therapeutics to humans and animals, whether or not we know their larger biological implications.

But the larger societal implications are a bit easier to forecast. The techniques of gene silencing, already well known to biologists, are just as easily knowable to anyone who's motivated to find out. The tools and molecules that are used for the procedure are available in pretty much any university or commercial laboratory. And it is remarkably cheap. Commercial firms are already selling sets of these short, custom RNA sequences, to silence any gene you'd like with more than a 95 percent success rate, for as little as $380.[8] With this in mind, you don't have to look far or think very hard to imagine what kinds of hazards might result from us mucking around with this powerful mechanism before we understand all of the ramifications.

For example, a research paper was published in 2004 that detailed how to silence the genes that control DNA repair functions in human cells.[9] The method the authors developed had been successful in keeping tumor cells from repairing themselves, and it was suggested in the article that the method might be used to make tumor cells more responsive to radiation therapy — that is, it could make the cells easier to kill. That would be an incredible boon for cancer patients, and a remarkable technological achievement. But our ignorance of the larger biological effects and implications notwithstanding, if this technique could make tumor cells easier to kill with radiation, there's no reason someone with a less noble goal in mind couldn't synthesize a small interfering RNA that blocked the cell repair function in normal cells and slip it into a common food source or the water supply. Such a molecular "dirty bomb," as Heinemann called it, could turn any organism that ingested or absorbed it into a kind of mutation factory. All the victim would have to do is go outside into some strong sunshine or breathe some toxic fumes or hang around in some other environment where cell damage would ensue. Imagine, for example, the damage if an important crop somehow had its radiation repair mechanisms silenced.

How realistic are these kinds of scenarios? While genetic engineering at this level of sophistication may be out of the range of ordinary people on the street, it's hardly in the realm of the esoteric. RNA interference is now a standard laboratory procedure for discovering new genes. Beyond the ability to create the interfering RNA itself, there's the further problem of detection. It's beyond the capabilities of today's technology to quickly or efficiently detect the many thousands of tiny RNA molecules capable of initiating gene silencing, either accidentally or on purpose.

Most computer analysis tools in genomics recognize only the sequences that are much larger than the tiny double-stranded RNAs that trigger gene silencing. Even if someone wanted to look for these RNAs, there aren't yet any standardized tools sensitive enough to detect them, whether they occurred naturally or were put there on purpose. What's more, researchers are just starting to realize how many genes may be responsive to silencing.[10] One estimate says that there could be as many as 60,000 small double-stranded RNA molecules per cell, all busily silencing their targeted genes as part of a cell's normal operations.[11] How could anyone routinely screen container-loads of corn for one tiny foreign sequence of RNA?

"Think about bacteria," said Heinemann. "There are already bacteria in most of the food that we eat. Some huge percentage haven't even been named. If someone were to modify a soil bacterium with one of these interfering RNAs, what would we look for? No one is looking for genes that small. This would be unfindable."

It's also incredibly difficult even under the best of circumstances to diagnose an ailment whose symptoms are sub-acute, let alone track down its cause. In the case of transgenic dairy sabotage, said Heinemann, "the cows' milk production would simply go down a bit. They wouldn't be writhing around out in the field. That's why GMOs are different. Their effects can be subtle."

And that's why Heinemann says RNA interference is the perfect mechanism for industrial espionage. "We already know that we can transmit these RNAs through food into the animal who eats them," he said. "And there are lots of very subtle ways to convert commercial GMOs into something that carries these sequences. You can put them into an organism so it has no visible effect. The cows are just producing less milk. And the beauty of it from a criminal perspective is that the best way to do it is the easiest way — use the most common component of a commercial transgenic, the cauliflower virus promoter, and you'll be hiding them in plain sight."

≡

I found Heinemann's perspective on this troubling, but in the removed way that one usually thinks of things like terrorist attacks: it'll never happen here, not to me or to us. So I asked him if there was any way that gene silencing could be triggered accidentally in transgenic organisms. And the unfortunate answer was yes.

"Actually these small RNA molecules can be produced *because* of transgenes," said Heinemann. Every transgenic cell has added a new sequence of DNA to the organism's genome. When it starts to synthesize protein in its new home, the DNA, in turn, "produces a new class of small RNA inside the genome that no one ever looks for. And every one of them could have a subtle effect on everything eating it — earthworms, animals, insects, all the way up to humans. The only way you'd ever know one was present is by looking for that specific, tiny sequence of RNA."

That's why in every safety assessment of commercial transgenics he's done, Heinemann always asks companies to identify the novel RNA molecules that result from their specific transgenic insertion. But since his assessments have been part of public consultations and not the formal risk analysis required by government regulators, he doesn't know whether the companies are actually required to submit that information — or whether the regulators would know what to do with it if they received it.

Heinemann presented a good reason to hope they learn.

"Some people study earthworms because they're key players in the agricultural ecosystem," he said. "So let's say that in transgenic corn, the unique RNA sequence that's been produced doesn't 'do' RNA interference — it doesn't find a match that allows the double strand to form and silence anything in the corn genome. But let's also say that in the worm genome, there *is* a match for this specific RNA, and it's a match to an important gene. That means the worms eating the corn in the field could be significantly altered by whatever gene was silenced, or they could die."

Dead worms could be the start of an ecological disaster. By eating and expelling organic matter, worms improve both the nutritional content of the soil and its physical structure. In the case of mass wormicide in a corn field, the corn would likely be the immediate suspect. But what about effects resulting from a more subtle alteration, via RNA?

"Whatever gene was silenced might decrease the worms' appetites just a little, and as a result the amount of time it takes for them to turn over the soil might go from three months to six months," said Heinemann. "We're tied to an annual food production cycle, so anything that slows down that cycle even incrementally affects our ability to accumulate food. The rate of turnover of soil nutrients is an important factor in the context of food production."

The effects of this type of gene silencing could vary depending on a number of different factors — the geography and climate of the area, for example, might present the worms with different kinds of environmental stresses that they'd have to respond to in their altered state.

And then there's the issue of what happens when traditional genetic engineering meets gene silencing in the same organism.

"It's already known that some *Bt* toxins can have an effect on worms," said Heinemann. One study, for example, showed that earthworms fed *Bt* corn material lost nearly one-fifth of their initial weight by the end of the test, while those fed non-*Bt* corn gained 4 percent of their initial weight.[12] "One concern is that in combination with a couple of other innovations, like gene silencing, maybe we're making new worm phenotypes," he added. "Maybe the worms don't die, but if there's some new stressor in their environment — a drought, too much rain, something that alters the chemical composition of the soil — it could yield a combination of small effects."

We're already dealing with this problem in the form of pollution. Environmental pollution has been found to trigger all kinds of bizarre animal behavior, including changing the social and mating behaviors of a wide variety of species.[13]

But maybe that small RNA doesn't find a match in the worm, either. Instead, maybe it matches something in the human genome. Maybe you've been eating transgenic corn or soy for a few months or a few years, and by the time you're 35, you're more tired than you were at 30, or you're 20 pounds heavier, or you can't digest certain foods the way you used to.

"Maybe that's normal, but maybe it's not," said Heinemann. "How would you diagnose that? Or maybe there's a few more babies with fragile X syndrome.* Or a few more people who are contracting irritable bowel syndrome. Just a few, and their symptoms were mild. How would you track that?" The short answer is, you wouldn't.

In the laboratory, researchers can significantly alter nematode worms by feeding them custom RNA-laced bacteria so they produce these small, silencing molecules. The RNAs are taken up into their digestive tracts, distributed throughout their systems, and silence whatever DNA they're specific to. This type of research is now routine as a way to study the developmental patterns of worms, but "outside the lab, this could easily be done accidentally, or on purpose," said Heinemann. "The only way you could even come close to predicting the effect is if you know every RNA that is being made by every transgene and compare the sequence of those RNAs to every known important gene in any organism important to you, such as worms, people, butterflies, whatever."

Even then, what would that tell us? Not much, given how much we don't know.

Some will say that Heinemann is overstating the problem. Genes move between species all the time, says the counter-

---

* Fragile X syndrome is a genetic disorder caused by a genetic mutation on the X chromosome. It's the most common cause of inherited mental impairment. According to the Fragile X Research Foundation, the mutation is found in 1 out of every 2,000 males and 1 out of every 4,000 females.

argument. Living things are constantly bombarded with new RNAs anyway. Every time our bodies take on a virus or a parasite, this could happen. Every time an alien grain of pollen lands on a plant, or an animal eats a dicey bit of food or gets bitten by a mosquito, it could happen. Therefore the risk of randomly triggering a gene-silencing "event" is not a risk that's unique to genetic engineering.

Quite true. However, the *random* occurrence of a single virus infecting a single person, plant or animal — a single gene transfer happening once, somewhere on the globe — isn't the situation we find ourselves in today with industrial-strength genetic engineering. "The difference is that in commercial transgenics, we've made sure this new RNA is present in 100 percent of the crops of anyone who plants that seed," said Heinemann. "It's concentrated in very, very small places in the genome, and it can potentially reproduce. So we go from maybe a bad RNA virus being reproduced by chance in a corn plant by genes moving across species, to an RNA molecule that changes worm digestion, or something like that, that's concentrated in every cell in every plant that contains *Bt* or some other commercial transgene that's covering 80 million hectares of arable land on the planet."

≡

Leaving aside the discovery of RNA interference, the dearth of true, in vivo feeding studies — feeding real transgenic food to real humans and animals in order to test the possible health consequences — is shocking. In 2003, two European biologists reviewed the published studies and found that only 10 such studies had been done, even after more than a decade's worth of commercial transgenic food has been consumed by millions of people and animals. "It would seem apparent that GM food regulation is currently based on a series of extremely insufficient guidelines," the authors wrote in their conclusion. The studies they reviewed showed that in several experiments animals did in fact have reactions to eating transgenic food — including changes in their stomach linings, their body weight and the weight of internal organs. These changes weren't present in animals fed conventional food.[14]

Yet a Biotechnology Industry Organization (BIO) background paper on food safety states that transgenic food is safe because it is "extremely unlikely" that DNA would survive digestion.[15] Therefore it would be extremely rare for transgenes to have the opportunity to move into living cells to make pests of themselves. But this is clearly and obviously untrue, and we know it by our own experience, if we think about it for a minute.

"Saying that DNA survives digestion is not a radical statement," said David Thaler, an associate professor in the Sackler Laboratory of Molecular Genetics and Informatics at Rockefeller University, who has been investigating the interactions between food DNA and the human digestive tract. "We normally excrete some undigested fragments of food we eat in a quite normal way. It would be surprising if just the DNA were destroyed and yet you can make out the form of, e.g., a piece of carrot. Some fruits are spread because birds and/or other animals eat them and excrete the seeds, still alive and ready to germinate."

Indeed, in 1992, the year that transgenic foods were approved for sale, researchers studying Italian bears in the wild published a report showing that they could pinpoint the exact plant variety the bears ate by doing DNA tests on their excrement.[16]

There can be little debate that DNA often survives the digestion process as part of the normal function of an organism. But as Thaler put it, "just being in the gut, then excreted is different than being biologically active in the gut." Of greater concern from the perspective of risk, and also "extremely unlikely" according to BIO's food safety paper, is the possibility that the transgene would not just pass through the intestines, but would actively breach the defenses of microbes in the digestive tract, or human gut cells themselves, and transform these cells — to some unknown effect.

Yet that, too, has been proven: a study published in 2002, commissioned by the UK's Food Standards Agency, showed for the first time that bacteria in the human digestive tract could take up transgenic DNA. Granted, the amount was barely detectable and was found only in the stools of study participants whose colons had been previously removed.[17] But even in the healthy participants, the transgenic DNA survived its journey through stomach acid and made it to the small bowel — which also is host to plenty of bacteria.

Beyond microbes, can DNA from food cross over from the digestive tract into the blood? Yes, Thaler said, pointing to pioneering studies by Walter Doerfler and his colleagues at the Institut fur Genetik in Germany. There, they've shown in mice that a fraction of pure DNA, when ingested, is indeed subsequently found in the blood.[18] Other animal studies and one human study (of healthy humans, this time) show similar findings.[19,20]

What do all these studies mean? Clearly the "science-based" reasons for transgenic food safety, given to us by the biotech industry and the FDA, are on pretty shaky ground scientifically. Much more research has to be done on DNA in the digestive system before anyone can say for sure what is happening, or might be happening, to the tons of transgenic DNA that humans and animals have ingested. Very little DNA would have to survive the digestive process to be what Thaler calls "biologically consequential" from either a health or an evolutionary perspective, or both.

CHAPTER 8.
THE PROMISE OF TRANSGENICS

These dark-side scenarios about transgenics and RNA interference gone awry may seem needlessly alarmist. Yet the facts on the ground remain: the defense of transgenics as risk-free continues to be based almost entirely on false premises about the precision and predictability of genetic engineering, with no reconsideration of the risks posed by the largely unknown interrelationships between genes, genomes and powerful, newly discovered mechanisms like RNA interference.

This is the point in the narrative where some people are shifting irritably in their reading chairs. "Everything new is risky," they might say. "So what? Just think where we'd be if we'd stifled all the progress of the past hundred years because we were afraid of what might happen!"

Of course. There is no such thing as risk-free living, with or without genetic engineering. Progress has never been risk-free. Even the most comprehensive and thoughtful risk analysis won't be able to anticipate every consequence from every new discovery or technology before we decide to proceed. Biotechnology's promised benefits are undeniable: the genetic alteration that might make way for a pig organ to work in a human being who might otherwise die, or a mosquito that could possibly stop the death march of malaria — well, it's difficult to look at the lives that could be saved and say, "It isn't worth the risk."

But how do we know it *is* worth it? Why, exactly, are we so unwilling to ask that question — and demand a rigorous answer? Why aren't our scientists and regulatory officials skeptical of these promises, and why aren't they as demanding of evidence to support claims of benefit for transgenics as they are of "scientific" proof of risk?

We may not have commercial-grade transplantable pig organs just yet, or malaria-proof mosquitoes, but we have lots and lots of transgenic crops. After more than 10 years in the market, shouldn't these products, at least, have racked up a list of benefits as demonstrable and credible as recombinant DNA's contributions to health and the environment?

You would think so. You would think that by now the promise of transgenic food crops — the most widely commercialized product of biotechnology, grown and consumed by people all around the world — would have either come to pass or not. The benefits should be obvious, or at least knowable by now in some measurable, verifiable way. Yet the controversies rage on: about food safety, about environmental and health risks, about the social and economic benefits of these products and processes.

The question of proving benefit came up during a conversation I had with Kim Waddell, a biologist and former study director at the U.S. National Academies. Waddell spent four years surveying the landscape of transgenic organisms from a unique perch: as the human nexus for three separate studies on transgenics, including the ones on animal biotech and biological confinement that I mentioned in Chapter 2, as well as another on the environmental effects of transgenic plants.

Each study in its own way attempted to anticipate the unintended consequences of deploying transgenic organisms in the field, and each netted Waddell a hefty dose of political grief before its release. Because of the close ties that most university scientists and all U.S. regulatory agencies have with the biotech industry, publicizing independent research on transgenics has often been a political sticky wicket; the ones Waddell shepherded through the Academies were no exception. Reports that are both faithful to the science and don't invoke a fatal antibody response from industry or regulators are rare and require far more steely diplomacy than you might imagine. After four years, Waddell finally wearied of keeping politics from watering down the reports before they were published or presented.

"By the time I left the Academies in 2004, the only project I could possibly have gotten funded was a report on the benefits of biotech," said Waddell. "You know, something that started out with the premise, 'We've seen enough of the risks of genetic engineering, now let's look at the benefits.'"

The academics who had been involved in Waddell's studies — many of them with strong industry ties — were pleased with the prospect of such a study. "I wanted to do it, too," said Waddell. "After this much time in the field, I thought there should be enough data to verify the benefits, to compare the promises with what's really happened. If it's real, we absolutely need to share that with the world." So the call for papers went out, and an industry biologist with the intriguing title of "director of regulatory science" presented Waddell with a fat folder of documents detailing the claims of benefit from the industry's perspective.

"I told him, 'Of course, you know that we'll have to have an independent review of the data for these claims to make sure they're scientifically valid, just like we did with the risks,'" said Waddell. "And — well, that was pretty much the last I heard from anyone on the subject of a study on benefits."

How could trained scientists, even if they do work for a biotech company, justify this kind of behavior? Why would they expect they could get such a report published by the most highly regarded scientific body in the U.S. without subjecting their claims to the same standards of scrutiny and verification they'd been demanding about risk? Unfortunately, it's a rhetorical question. No one who understands the politics that drive biotechnology, including Waddell, has ever expected to see a truly independent study of the benefits of transgenics. Because in the U.S., at least, it's exceedingly rare to see anyone taken seriously who dares to challenge biotech's claims.

This state of affairs was distressing to at least one senior government regulator in the U.S.: Dan Glickman, who headed the USDA in the Clinton administration (and who now is president and CEO of the Motion Picture Association of America). Glickman's experience at the USDA speaks directly to the second half of the benefits question: To whom do the benefits accrue?

By virtue of his position, Glickman was responsible for evaluating the safety of several classes of transgenics. During his tenure at the USDA, he pushed for (and funded one of) the reports that Waddell directed. He also founded and provided the initial funding for a standing advisory group on agricultural biotech in the National Academies. By the time he left office, he'd seen enough to deliver the following indictment of the U.S. attitude toward biotech during an interview with Bill Lambrecht, a reporter for the *St. Louis Post-Dispatch* and author of *Dinner at the New Gene Café*:[1]

"What I saw generically on the pro-biotech side was the attitude that the technology was good, and that it was almost immoral to say that it wasn't good, because it was going to solve the problems of the human race and feed the hungry and clothe the naked," Glickman told Lambrecht. "And there was a lot of money that had been invested in this, and if you're against it, you're Luddites, you're stupid.

"That, frankly, was the side our government was on. Without thinking, we had basically taken this issue as a trade issue and [our position was that] they, whoever 'they' were, wanted to keep our product out of their market," he added. "You felt like you were almost an alien, disloyal, by trying to present an open-minded view on some of the issues being raised. So I pretty much spouted the rhetoric that everybody else around here spouted; it was written into my speeches."[2]

This market-driven rhetoric about the promise of genetic engineering is all that most people in the U.S. have been exposed to. The benefits of biotechnology are treated as self-evident, and those who challenge them are, as Glickman noted, generally met with hot scorn or derision. But Waddell had the right idea. A truly independent examination of the benefits of genetic engineering would be the only way we'd ever get anyone to undertake an honest re-evaluation of the risks.

There's no shortage of propaganda on either side of the debate. At the time this book was written, independently reviewed data on benefits had only just begun to trickle in via the scientific literature. In what follows, I've tried to present the biotech industry's claims as made to its constituents — primarily the ones made to its customers (the farmers), but also to some degree what was promised to the public by the industry's public relations campaigns — and then to present the challenges to those claims, based on the most credible data I could find. It's not an optimal solution, but perhaps the result will be sufficiently irritating to all parties that some organization will support a truly independent review of biotech's benefits.

≡

To begin, as Glickman noted, the initial appeal that the biotech industry made to the public was that genetically modified food would provide the solution to world hunger — that the transgenic crops would allow farmers all around the world, especially in developing countries, to safely grow more food to feed themselves and sell the rest to others. And while it may be unseemly to stick a pin in such a big and apparently worthwhile balloon right off the bat, by most accounts this claim has always been something of a red herring.

Many knowledgeable organizations and institutions that study and gather data on world hunger and poverty, including the United Nations, maintain that the reasons for global hunger have nothing to do with the amount of food in the world. In 2002, according to a report sponsored by three international agriculture organizations, world agriculture produced 17 percent more food calories per person that year than it did in the 30 years previous, despite a 70 percent population increase. "The food is there," the report bluntly stated. "The existence of 780 million chronically hungry people in the developing world today shows that there is something fundamentally wrong in the distribution of food and the resources with which to access it."[3] In other words, people who are hungry don't have sufficient land to grow, or income to purchase, enough food.

Beyond the world of agricultural biotech, political corruption and unequal distribution of wealth are the two most often cited reasons for hunger, and an infinity of transgenic food doesn't address either issue. However, the *over*production of food — particularly the $1 billion spent per day to subsidize the farmers in wealthy countries to grow too much of it — is often overlooked as a source of chronic poverty and hunger. This fact is clearly relevant to whether or not transgenic crops are "feeding the world," as one biotech company's advertisement phrased it.

In essence, big, rich governments, such as those in the U.S., the European Union, Japan and China, subsidize their farmers, paying them to grow too much food. This in turn drives down the price of commodity crops in the global market — a consequence that has devastating effects in developing countries. As *The New York Times* said in a 2003 series on hunger and poverty, "In recent years, American farmers have been able to dump cotton, wheat, rice, corn and other products on world markets at prices that do not begin to cover their cost of production. Poor nations' farmers find they cannot compete with subsidized products, even within their own countries."[4] This makes poor farmers even poorer, and hungrier.

So until the geopolitical situation has righted itself, the stated benefit of biotech sating the world's hungry is off the table, so to speak. While someday we are likely to face a true shortage of food in the world, for now food scarcity is not an issue that's being addressed by transgenic crops — in fact, in light of how some farm subsidies work and the way biotech products are sold, transgenics may actually be making the situation worse.

The real market appeal of transgenics is a much simpler proposition. In the case of insect-killing transgenics like *Bt*, the benefit to farmers is that they should be able to spray far less insecticide or even none at all in order to kill the specific insects targeted by the *Bt* toxin. This means farmers don't have to buy or expose themselves to these costly and dangerous chemicals.

With herbicide-tolerant crops like Roundup Ready soybeans, the benefit is flipped. Because the plants produce the transgenic enzyme that can tolerate Monsanto's Roundup weed killer, farmers can spray entire fields with abandon, killing weeds wholesale but leaving the crops standing. Another big selling point for herbicide-tolerant plants like Roundup Ready soy is that farmers don't have to till the soil in order to plant new seed. Traditionally, farmers till their fields to get a head start against the weeds — breaking up and burying the seed beds that established themselves during the previous growing season. But loosening the soil by tilling can lead to soil erosion. Erosion pollutes water sources with runoff, changes the composition of the remaining soil and, adding insult to injury, shrinks the amount of land that's suitable for planting. A "no-till" crop is generally thought of as more ecologically sound, especially when it's part of a larger land-use strategy that includes practices like crop rotation.

Taken together, the promised benefits of today's most popular transgenic crops (*Bt* and Roundup Ready) were to make farming less chemically dependent and less labor-intensive, thus more profitable for farmers. This has been a strong inducement for farmers around the world to buy and plant transgenic crops. It's the farmers, not those who eat the food they grow, who were to be the primary beneficiaries of transgenics.

≡

Have these claims held up over time? One of the few scientists trying to answer this question with data is Charles Benbrook. An agricultural economist, Benbrook spent nearly 20 years in Washington D.C. in a variety of senior advisory or staff roles, from the Carter White House to the Committee on Agriculture in the U.S. House of Representatives. In 1984 he was recruited to run the Board on Agriculture at the National Academies. Now at the helm of a consultancy that's focused on sustainable agriculture, Benbrook has researched and published several technical analyses that are critical of the impacts of transgenic crops on farming and farmers. The field data he has seen so far doesn't tell the same glowing story that farmers are told by the commercial purveyors of transgenics.

In a 2001 analysis of the "farm level" economic impacts of *Bt* corn, using data he purchased from Doane Marketing Research, an agricultural market research firm founded more than 80 years ago in the heart of the U.S. farm belt, Benbrook discovered that *Bt* corn yielded up to 30 extra bushels per acre.[5] But he also noted that to access this "remarkable technological achievement," as he called it, American farmers were paying a premium for the seed — 30 to 35 percent more than the cost of successful conventional varieties. By his calculations, the introduction of the *Bt* variety represented the highest per-acre seed expenditure in history linked to a single new trait in corn hybrids.*

Of course, a price hike that steep wouldn't matter if the money saved on labor and insecticide covered the extra expenditure. But did it? Not according to Benbrook's report. "*Bt* corn has not come close to keeping pace in returning value to the farmer," he wrote. "In fact, the economic benefits of *Bt* corn have not even covered added expenses when averaged across all acres planted over the last six crop years."

Given Benbrook's previous roles in a government that officially championed biotechnology, I asked him if he'd had some kind of conversion experience since leaving Washington.

"No, it's not my life's mission to run a one-man truth squad for the biotech industry," said Benbrook, in a telephone conversation. "I keep getting asked to do these studies because I know how to use the government databases to make those specific kinds of calculations. But I do think the biotech industry has

*The data purchased by Benbrook Consulting Services included acres planted, units planted (a "unit" is a bag of seed containing about 80,000 kernels, enough to plant 2.5 to 3 acres), average retail prices per unit, average discounts, and net prices (retail prices minus discounts).

really been outrageous in many of its claims of benefits, and in particular its claims of reducing pesticide use with GMOs. As somebody who has spent a professional career trying to bring reason and fact and logic to complex and often contentious public policy issues, I just felt that it was very important to set the record as straight as it possibly can be in terms of lack of data about benefits."

More recently, in 2004, Benbrook published a report on the first nine years of transgenic crops and pesticide use in the U.S.[6] Since the term "pesticide" includes both weed killers and insecticides, his analysis included Roundup Ready soybeans as well as *Bt* crops. Using statistics from the USDA and consulting with pest management, insect and weed science specialists at several universities, Benbrook's analysis showed that the news about herbicide-tolerant crops was even less favorable than his financial analysis of *Bt* corn.

Proponents have claimed repeatedly over the past decade that transgenic crops reduce overall pesticide use, but apparently this has not been true since 1999. *Bt* crops did in fact reduce insecticide use in the U.S. by about 15.6 million pounds over the nine years included in Benbrook's study. But this gain has been more than offset by the increase in herbicides due to the proliferation of Roundup Ready plants. Crops that are genetically tolerant of weed killers like Roundup require more herbicide, because farmers will dose an entire field with herbicide and not just the weeds. As a result, farmers using Roundup Ready plants have increased herbicide use by 138 million pounds since 1996. Taking into account the reduced insecticide use for *Bt* crops, this still totes up to a whopping 122 million-pound increase in pesticide use overall.

Not surprisingly, the volume is increasing. Recall the earlier discussion about the need to "bioconfine" plants that have been engineered with herbicide tolerance. Weed scientists have been sounding the warning since the first transgenic seed was sown that relying too much on crops with genetic tolerance to herbicides would trigger changes in weed communities, much like the overuse of antibiotics has triggered antibiotic resistance in microbes. And indeed it has. This growing tolerance in weeds in turn has forced farmers to apply more weed killer, and/or apply it more frequently, with poorer and poorer results.

By 2004, weeds resistant to herbicides had been showing up for three or four years around Roundup Ready crops "and appear to be accelerating," according to Benbrook's report. As a result, and for the foreseeable future, herbicide tolerance "will increase pesticide use more than *Bt* transgenic crops reduce it."[7]

And how about this for irony? While resistance is bringing about increased herbicide use on transgenic crops, said Benbrook, regulators and agrichemical innovations have been successfully edging herbicide usage *downward* on conventional varieties.

≡

The irony continues to be lost on transgenic soybean growers in Argentina — a country that, as Benbrook noted, "has adopted GM technology more rapidly and more radically than any other country in the world." According to statistics from the Argentinean government, in the 2003-04 growing season a record 14.5 million hectares were sown with soy, compared with just 37,700 hectares in 1971.[8] Farmers drastically reduced the amount of most of the other crops they used to grow (including corn, wheat, lentils and rice) in order to plant soy, which was in great demand and fetching a high price as animal feed. Approximately 99 percent of that is transgenic, making Argentina the second-largest transgenic soy producer in the world behind the U.S.[9]

In the early years of Argentina's adoption of transgenic soy, the environmental benefits seemed to match the economic boon. Since the late 1980s, about half of the country's largest and most fertile farming region had been suffering from soil erosion so severe that yields on the affected land had fallen by at least a third, according to the country's National Institute of Agricultural Technology (INTA). Farmers were experimenting with no-till methods, but with no plowing, the weeds were taking over, and Roundup Ready plants seemed the perfect solution. Soil erosion declined as more Roundup Ready soy was planted and farmers switched herbicides to Roundup, considered to be one of the least toxic herbicides available.

By charging a steep export tax on soy, the Argentinean government benefited as richly as the farmers. Transgenic soy became the government's number-one source of income during a time when the country's economy was literally collapsing. That was possible in large part because Monsanto was selling Roundup Ready seeds to Argentinean farmers at a fraction of the cost they sold for elsewhere; it hadn't made the same lucrative "licensing" deal in Argentina that it had made with farmers in other countries. Also, the Roundup herbicide itself had never been protected by a patent in Argentina; as a result, several other chemical companies had commercialized it under different brand names and forced the price to 25 percent of what it had been in the mid-1990s.[10] The cost to cultivate Roundup Ready soy was so low that even when soybean prices started to decline as global supply increased (in part because of Argentina's pell-mell rush into the market), and even when weeds were starting to grow tolerant of glyphosate and require more frequent and heavier application, Argentina and its soy farmers continued to make money hand over fist on the crop.

But transgenic soy farmers in the region sprayed so much glyphosate that peasants on neighboring non-Roundup Ready farms successfully took the soy farmers to court to stop the toxic clouds from blowing onto their land and damaging their crops. The judicial relief was only temporary, though. As soon as new renters took over the transgenic crop, the dousing began anew. In addition to crop damage, neighboring farmers and their children have developed nausea, vomiting, diarrhea, stomach pains, skin lesions, allergies and eye irritation from exposure to the herbicide.[11] One article reported that after the land had been sprayed with glyphosate, a neighbor's "chickens and pigs died, and sows and nanny goats gave birth to dead or deformed young. Months later banana trees were deformed and stunted and were still not bearing edible fruit."[12]

The soy crop, however, was in fine shape.

But perhaps not for long. While humans and animals are starting to react to massive doses of glyphosate, weeds themselves aren't responding to it as they once did. Glyphosate-tolerant weeds, once uncommon, are becoming more abundant,[13] and agronomists from a local INTA office have documented farmers using higher concentrations of glyphosate to control them. Some scientists are warning that it's only a matter of time before glyphosate resistance is transferred to other weed species beyond the soybean fields, potentially creating a race of death-defying "superweeds."

In fact, in the summer of 2005 scientists were arguing over the finding of a British research team that just such a superweed already had been created: a hybrid of an herbicide-tolerant oilseed rape plant and charlock, a related weed.[14]

There's also growing evidence that the soil itself is changing as a result — that the massive applications of glyphosate are killing off the bacteria that keep soil "alive" and healthy and that keep other diseases at bay.

Adolfo Boy, an agronomist from Grupo de Reflexión Rural, an Argentinean group opposed to transgenic farming, claims that some of Argentina's legendarily fertile soil is being rendered inert as a result of the chemical onslaught. So spent is the soil that "in some farms the dead vegetation even has to be brushed off the land," Boy told a journalist for *New Scientist* magazine in 2004.[15] As a result, new pests, such as slugs, snails and fungi, are moving in — which, of course, need to be controlled with pesticides. One newly arrived fungus affecting soy can cause crop losses of up to 80 percent, and the fungicides to kill the disease can cost as much as $50 per hectare to apply.[16]

Benbrook said that U.S. researchers have begun to study the relationship between glyphosate use, a growing onset of fungus in soybean crops, no-till methods and what's known as "continuous cropping" — growing the same crop in the same fields over and over without rotation. "On a scale of one to 10 of how to create a pathogen problem, no-till will get you into the 3 or 4 range," he said in an interview. "But with continuous cropping, you're up to 6 or 7. Once you're up in that high of a risk category, you'll get some disease almost every year. And there's a threshold effect. The effects can be subtle to dramatic, but even effects that farmers can't observe yet are probably impacting yields by half a bushel, or a bushel or two, a year."

The need to kill the "volunteer" transgenic soy plants that have grown from seeds dropped during harvest has presented farmers with another pesky problem: how do you kill an herbicide-tolerant plant? Chemical companies — one of which uses the slogan "Soya is a weed!" — are trying to sell farmers on a mixture of paraquat and atrazine as one solution.[17] Paraquat is a defoliant, toxic to humans as well as plants, that earned infamy in the 1970s when it was used by U.S. drug authorities to destroy Mexican marijuana fields. Low doses of atrazine, a widely used herbicide, have been shown to turn some amphibians into hermaphrodites and deform their limbs.[18]

So much for transgenics decreasing the use of dangerous pesticides.

Argentinean agriculture officials tend to blame the country's growing environmental problem on the monoculture that has sprung up around Roundup Ready soy, rather than the transgenic variety itself. But whatever the culprit, a December 2003 report from INTA warned that if nothing was done about the effects of soy on the environment, a decline in yield was inevitable and the country's "stock of natural resources would suffer a (possibly irreversible) degradation both in quantity and quality."[19]

"Argentina faces big agronomic problems that it has neither the resources nor the expertise to solve," said Benbrook in a 2004 report on the situation. "Based on the current use of Roundup Ready, I don't think its agriculture is sustainable for more than another couple of years."[20]

The impact of transgenic soy farming has been as radical and destructive to the people of Argentina as it has been to the country's once lush farmlands. A case study on soy in Argentina, stuffed with government statistics and citations from university studies, details the social devastation that has resulted from Argentina's unchecked race into the global soybean market. As the rent skyrocketed for land that could grow soy, peasants were forced off farms and into the cities, which could not support them. The loss of some traditional crops and the drop in production of others meant importing them at higher cost, making it harder for the poor to purchase any traditional food.

"Those who cannot pay must survive on handouts, leftovers and produce that cannot be sold," according to the agronomist Boy.[21] Or they must survive on soybeans, which are not recommended as a primary protein source for humans, especially not for children. As a result, many Argentineans are going hungry.

"People could not believe that there was hunger in a country that produced so much food," said the report's authors. "What they did not realize was that the country no longer produced food, but animal feed for livestock in far-off countries."[22]

≡

What does all this have to do with risk? Some will say that what happened in Argentina can't be blamed on transgenic soy, but instead is the fault of the farmers and their government who rushed to cash in without thinking through the consequences. If the transgenic "technology" had been used properly, the argument goes — if crops had been rotated, less land had been planted with soy, etc. — surely at least some of the promised long-term benefits would have come to pass.

Possibly. But that's not what happened. *A credible risk analysis cannot claim a technology is safe unless it includes all the ways that people can use it — including how they can misuse it.* This is no longer new information! We have known this truth for many decades, at the very least since operator error was judged responsible for the Three Mile Island accident in 1979. Risk analysis isn't a crystal ball into the future. But when those who represent a government or an industry are trying to break into a market with a new and untested technology, and they don't include important variables like human factors in the analysis, then they should be held accountable for whatever disaster or havoc that may ensue.

In Argentina, the government wasn't required to think or care about the long-term effects of a soy monoculture on its land, nor the attendant long-term effects on its economy and the health and welfare of its citizens. So its farmers did as they pleased, and both farmers and government collected the bounty. During Monsanto's first several years of doing business in Argentina, before it abandoned the market in 2004, the company was welcomed with open arms by farmers and the government alike — even though any professional agricultural biologist or economist employed by Monsanto would have known that producing a commodity in a system of near monoculture could be disastrous.

"Monoculture is not good for the soils or for biodiversity and the government should be encouraging farmers to return to crop rotation," said Carlos Senigalesi, who in 2005 was director of investigative projects at INTA. "But here everything is left to the market. Farmers have no proper guidance from the authorities. ... I think we must be the only country in the world where the authorities do not have a proper plan for agriculture but leave everything to market forces."[23]

As a result, what was once some of the most fertile land in the world may be laid to waste, and some experts are afraid it may not be reclaimable. Adding injury to insult, more people are poor and hungry in Argentina in 2005 than before the first transgenic soy seed was sown.

≡

Equally confounding situations are unfolding all around the world for another commercial transgenic crop that heralded great benefits for farmers: cotton plants engineered to produce *Bt*'s insecticide protein.

Bt cotton is one of the most controversial of all the commercial transgenics. Outside of the developing world, in the U.S. and Europe, for example, cotton hasn't engendered as much controversy as other transgenic crops because it's mistakenly not considered to be a food crop. But in addition to the use of cotton fiber for textiles, cotton seeds yield cakes for animal feed, as well as oil that's commonly used as a salad or cooking oil (Crisco, for example, is simply hydrogenated cottonseed oil that looks like lard). In fact, cottonseed oil production is third in volume only to soybean and corn oils.[24]

Apparently cotton is a popular food source for insects as well. According to one estimate, cotton is responsible for 25 percent of all insecticide use and more than 10 percent of all pesticide use on the planet, including some of the most hazardous. Because of the health hazards these chemicals present for cotton farmers, the benefits from *Bt* cotton cultivation were to be simple and threefold: One, reduce the use of pesticides to be sprayed on the crop. Two, reduce the cost of cultivation. And three, make more money through both of these cost reductions plus the resultant higher yields. Heavy promotion of these benefits by purveyors of *Bt* cotton prompted countries on many agrarian continents to begin growing *Bt* cotton commercially or conducting field trials — these countries included the United States, of course, as well as Australia, China, South Africa, Mali, India, Indonesia and others.

Although there are many stories told about how the promoters of *Bt* cotton have elbowed their way into various countries, one of the most openly contentious markets for transgenic crops has been India. Protests against the adoption of genetically modified food crops — some of them violent — have been an ongoing fact of life for nearly 10 years all across the country. But there's at least one good reason that the protesters haven't yet won over the country's cotton farmers: fully 50 percent of the pesticides used in India are applied to cotton.

Given the importance of the issue to its people, it's no surprise that the most detailed and credible field study I've seen challenging the benefits of *Bt* cotton came from India. Based on data gathered during the 2003-2004 growing season in the state of Andhra Pradesh, the study was funded by two local sustainable-agriculture groups and conducted by two agricultural scientists, one of whom is the former joint director of agriculture in Andhra Pradesh and the other of whom is a scientist at the Permaculture Institute of India.[25] The same team had studied and published reports on the performance of *Bt* cotton for the previous two seasons as well.

The scientists and their data collectors spent nine months documenting farmers' observations and the seeding rate and density of crops in the cotton fields of Andhra Pradesh. They recorded every farmer's cotton-related income and expenditures, as well as whether they grew *Bt* or non-*Bt* crops. In addition, every two weeks for the entire growing season, they recorded data on field operations — the use of fertilizers and pesticides, the status of crop and pest damage — in 164 fields of varying sizes from 28 villages. A team of scientists would also randomly visit fields in the villages that weren't otherwise included in the study in order to record those farmers' experiences. All of this was done every 14 days from the sowing of the cotton crop until it was harvested. The result was not a ringing endorsement of *Bt* cotton. According to the scientists' data:

- Farmers spent 230 percent more money for *Bt* seeds than non-*Bt* hybrid seeds.
- *Bt* cotton required 8 percent more total investment than non-*Bt* cotton.
- The reduction in pesticide consumption by *Bt* farmers was only 12 percent.
- Net profits from *Bt* were 9 percent less compared to profits from non-*Bt* hybrids.
- The cost-benefit ratio favored non-*Bt* hybrids.
- For small and medium farms, the yield difference between *Bt* and non-*Bt* was negligible.

Just as important as the economic data collected was the information farmers gave the scientists about the ways in which *Bt* cotton was behaving in their fields. After only two years of *Bt*, farmers said they were already seeing signs that major cotton pests were developing resistance to the *Bt* toxin — a development that would also be catastrophic to other important crops in the area that could serve as alternate hosts to cotton pests.[26] Farmers also told the researchers that *Bt* seemed to be infecting their soil with a special kind of root rot, and no other crops would grow in that same soil. The same problem was not found in non-*Bt* fields.

Monsanto, whose Bollgard cotton is the *Bt* variety of choice in India, had solicited its own survey, hiring the market research firm A.C. Nielsen to conduct it.[27] The Nielsen study was used as a sales tool around the Indian countryside; it presented the results of interviews with a random sample of cotton farmers throughout India. Compared to the local research scientists' approach, Nielsen's interview period was much briefer — about three weeks, instead of nine months — and was less detailed. Monsanto's study showed that *Bt* crops had far fewer bollworms than non-*Bt* crops, and the farmers had vastly greater savings on pesticides, much higher yields and a significant increase in net profit, where the local study showed a 9 percent loss.

Overall, while both were officially "quantitative" studies, they agreed about precisely nothing.

Several months after the Andhra Pradesh report was published in early 2005, *Bt* stories started cropping up in various outlets. The journal of the Indian Academy of Science published an article, from India's Central Institute of Cotton Research (CICR), that predicted how, when and under what circumstances cotton bollworms might develop resistance to *Bt* varieties. Reviews of its data produced some provocative articles in the scientific and mainstream press. A *Nature Biotechnology* news article surveying some of the more recent studies, including CICR's, drew fire from one of the study's authors who claimed his results had been misrepresented.[28]

"We strongly believe that the *Bt* technology is the best eco-friendly tool available for cotton pest management in India," wrote Keshav Kranthi of CICR's Crop Protection Division in response to the article.[29] But who's spinning whom? A footnote to Kranthi's article states that the CICR study was funded by Mahyco, Inc., Monsanto's seed partner, which in 2002 became the first Indian company to commercially grow and market *Bt* cotton, India's first transgenic crop.[30]

Indian farmers are well known to be experimenting with creating illegal hybrids of *Bt* using their own seeds. And so now, some unknown yet significant amount of *Bt* cotton is being grown all over India, to who knows what eventual effect. As Kranthi wrote in his protest to *Nature*, "Indian farmers have always overused good technologies to the point that they are rendered useless, as was the case with good varieties and insecticides." So why didn't India's regulators address these well-known human factors before approving the transgenic varieties? And "why weren't rigorous studies such as [Kranthi's] conducted earlier," as one biotech opponent asked, given the potential for abuses that could lead to resistance and monoculture?[31]

"We're now asking ourselves the same question," said a government entomologist.[32] Given that CICR's study was funded by a biotech company under the directive of the Indian government's Genetic Engineering Approval Committee, Indian citizens should probably be grateful for at least that much self-reflection.

The fact that transgenic crops like Roundup Ready soy and *Bt* cotton have become so firmly entrenched in the global food chain under such circumstances is sobering. Why haven't government regulators demanded proof that the benefits of transgenics outweigh the tremendous risks that farmers around the world have taken on faith?

CHAPTER 9.

THE TRICKY CALCULUS OF COST AND BENEFIT

When it comes to risk, "benefit" isn't just about promises. Benefit is the equal and opposite variable in a very important equation. We already know that some kinds of risk can be mathematically described and that risk calculations can have some degree of value under the proper circumstances. But if we don't want to repeat the mistakes made in Argentina and India (and other places where transgenes have contaminated traditional croplands), we also have to be very clear about benefit — not only how it's calculated, but how benefit manifests in the real world.

Virtually every strategic or official risk decision about technology that's made in a government regulatory agency or ministry, or in the corporate environment for that matter, is supported either formally or informally by a very tightly focused calculation called a "cost-benefit analysis."

The term *cost-benefit analysis* essentially describes its function: using some metric — usually financial, but not always — you draw up a balance sheet for the problem you're trying to solve. You calculate the probability of the benefits you hope to gain from an action or a process on one side, and calculate the cost, or risk, on the other. You tote up the columns and compare them. The calculation should show you whether the risk is worth the benefit — if the pain is worth the gain.

Applied to the right kinds of problems and used appropriately, cost-benefit analysis uses the laws of probability to anchor a decision in real outcomes. In the corporate world, it helps executives think in a structured way about the true cost of entering a market, moving the company in a new strategic direction or fixing various kinds of existing problems. Government regulators rely on cost-benefit analysis to decide whether the cost of regulating a product (like transgenics) or some kind of activity (like smoking in restaurants) is worth the benefit realized in protecting the public. The ways in which governments use cost-benefit analysis to make these decisions is largely responsible for determining which risky situations, products, or processes we're exposed to in our daily lives.

≡

There are two kinds of cost-benefit equations that are important to government regulators. One is the cost to government of creating and enforcing the regulation — the actual bureaucracy — balanced against the cost to the public if they were left exposed to a hazard. But the second kind of cost-benefit equation presents a far more tricky calculus and generates the greater share of today's controversies about regulating transgenics and other products of the biotech industry.

It's a government's unenviable job both to protect the public safety by regulating potential hazards and to promote its national industries and economic interests. Is it better to strictly regulate a new market so the public feels "safe," possibly forcing your best and brightest scientists and entrepreneurs to accept or create jobs elsewhere? Or should the risk burden be shifted onto the public — in essence, asking them to pay for the development of a new market in exchange for the benefits they'll receive?

This is far from a black-and-white situation; in fact, it's the paradox that virtually all governments face if they want to participate in the global economy. Commercial scientific innovations (new drugs, new forms of energy, new biotechnologies and so on) might bring both direct benefit to the public in terms of improving the quality of life and indirect benefit by stimulating economic growth. But they might present both direct and indirect risks as well. Innovations can turn out to be more hazardous than their creators could know, and if regulation (or lack of it) favors growth over protecting the public (sometimes from itself, as was true in Argentina), the promised benefit can come at great social and economic cost.

How a government manages this paradox — how openly and honestly it addresses the need to balance public risk and benefit with private risk and benefit, how it finds and cultivates the areas where they intersect — determines the degree to which both industry and the public invest their trust and their money in government decisions that require scientific expertise.[1] For this reason, knowing exactly what goes *into* the analysis, in the form of existing knowledge and assumptions and methods for defining the problem, is so important to cost-benefit calculations.

For example, if you were trying to responsibly evaluate the risk of a food produced by a radically new method, as the FDA was back in 1992, why would you completely eliminate from the discussion any evaluation of the method that produced it? Why *wouldn't* you evaluate the process before deciding whether or not it was relevant to safety? The Reagan administration's intervention into that process was at least as critical a risk factor as the science upon which the decision was (or wasn't) based. In fact, while the political conditions that led to substantial equivalence appear to be a particularly eye-popping corruption of the risk assessment process, most of us know about it only because the discovery documents got posted on the Internet. But rest assured that this kind of stacking the deck — this "gaming" of risk analyses — is not unusual. It happens all the time in the interstices between government regulators and all kinds of industries.

A recent example illustrates why, and how.

In March 2005, the EPA announced what it calls a "rule" or new regulation that specified how it would (or wouldn't, more to the point) limit mercury emissions from coal-burning power plants. As you'll recall from the earlier discussion of transgenic, mercury-aspirating trees, mercury exposure is extremely risky to human health. In a large lake, a single drop can make all the fish in it unsafe to eat, and mercury exposure has been traced to many serious health problems in adults and in unborn children.[2]

At the time the rule was announced, EPA officials declared that emission controls could not be more aggressive because the cost to industry — $750 million — far exceeded the public health payoff, which they estimated at $50 million a year. But a Harvard University study on the subject, paid for by the EPA, co-authored by an EPA scientist and peer-reviewed by two other EPA scientists, had reached a vastly different conclusion. The Harvard study concluded that the same $750 million could save nearly $5 billion a year — *100 times* the amount of the EPA's estimate of the public heath payoff — by decreasing the neurological and cardiac damage attributed to mercury poisoning.

As it turned out, the EPA's final analysis included only the effects of reducing mercury levels in freshwater fish, while the Harvard study acknowledged a more relevant fact: that most of the fish Americans eat comes from *oceans.* "Some very large share of mercury exposure comes from tuna," James Hammitt, director of the Harvard Center for Risk Analysis and co-author of the Harvard study, told *The Washington Post* at the time the ruling was announced. "And while it's true that our power plants have less effect on tuna than on [freshwater fish], if you ignore the saltwater pathway you'll miss a lot of the benefit."[3]

The EPA also greatly reduced the "cost" of cardiac damage in its risk equation, declaring that although mercury could indeed damage the heart, the harm might be offset by the known cardiac benefits of eating fish.[4] This statement was so cynical that it rendered me speechless at the time I heard it. How would someone even set up the experiment to test that theory? Well, at some point, it won't be possible, because there won't be any untainted fish left to make the comparison.

How the EPA used its statistics is the classic example of what you can do when the numbers in a cost-benefit analysis aren't turning out the way you want them to. You come up with a justification for leaving something out or putting something in, something that's at least semi-justifiable if someone catches it, and a dramatically different outcome ensues. In this case, Harvard's cost-benefit analysis estimated 100 times greater health benefits than the EPA's. It's hard to argue with numbers like that. So to keep "costs" down for the EPA's constituency — which in this case appeared to be the coal industry — it simply picked the facts that "proved" eliminating mercury emissions won't make that much difference, health-wise, thus demonstrating how little need there is to control them.[5]

≡

No matter what industry you're in, if you've got the nerve and the know-how, gaming an official cost-benefit analysis can be irresistible, a way to coax the future over to your side of the table every time, no matter what your "side" happens to be.

That is, unless there's some official means by which outsiders can review and challenge the assumptions behind your cost-benefit studies. Recall Paul Thurman's earlier exhortation about the critical importance of transparency for any kind of analysis that includes statistics and probability. Unless all the components of a cost-benefit analysis are made public and explicit, and there's a way to challenge its assumptions, there's truly no way to know what the results actually mean, including whether or not they're plausible or credible in the real world.

This being the case, we arrive at the question: How do you gauge risk, and what method do you use, when questions of cost and benefit are unfolding within an intensely charged context — a context where social and political influences press up hard against scientific uncertainties, like those that confront us with the fruits of commercial transgenics? I've not yet seen this question addressed in any scientific discussion or cost-benefit analysis of commercial transgenics.

Saving seed is one example that should demonstrate why the context of how we measure risk is so important.

Saving the seeds from the best plants at harvest to plant in the next growing season is possibly the most fundamental farming and agriculture practice since human beings taught themselves to purposely grow food. As farming became more commercialized, seed companies developed higher-yield hybrid varieties of various food crops like corn, sunflower and sorghum that lost a significant percentage of their growing advantage by the end of a single season. Farmers had much less incentive to save seed from these crops and instead bought fresh seed each year.

But wheat, soybeans and rice are self-pollinating crops, and even commercial varieties don't require that farmers purchase new seed each year. This is because the seed harvested from the crop is genetically identical to what was planted. Growers of these crops need to buy seed only once; they can keep and plant their own from year to year. Hybrid crops don't present this opportunity and thus have a kind of built-in biological intellectual-property protection for seed producers. But without a way to induce farmers to buy new seed each year, self-pollinating crops historically have been an unattractive investment for commercial companies.

Enter transgenic crops, which have been patented by biotech companies. Farmers who want to grow transgenic crops are required to sign grower agreements that prohibit farmers from saving seed to plant the following year. They are really only buying a license to "use" the seed for that growing season. It's much the same as when we "buy" a software package, but the legal reality is that we are simply paying the manufacturer a licensing fee for the right to use it under the conditions they decree. Similarly, soy farmers pay a "technology fee" of about $6.50 an acre each year. This fee is the premium that Charles Benbrook was talking about in his report on farm-level effects of transgenic crops.

When farmers sign one of these license agreements, they also grant the biotech company the right to come onto their land at any time and test their crops for the transgene. If any transgenic plants are found and the farmers haven't purchased transgenic seed for that growing season, they can be charged penalties that are as much as 120 times the original technology fee, plus they must pay the biotech company's legal fees — even if the plants were "volunteers" from last season's dropped seed.

If they refuse to pay, the company can sue, and very often it does. Since 1997, Monsanto has filed more than 90 lawsuits in 25 U.S. states against 147 farmers and 39 agriculture companies.[6] In a 2004 case, a Tennessee farmer was sued by Monsanto and sentenced to eight months in prison after he hid a truckload of harvested cotton seed for a friend. He has been ordered to pay Monsanto more than $1.7 million for his infraction.[7]

This injunction against saving seed is happening around the world. In order to trade with any country that belongs to the formidable World Trade Organization (WTO), you have to agree to abide by its rules about what it calls the "trade-related aspects of intellectual property rights," or TRIPS. The WTO, of which the United States is a powerful member, passed the TRIPS accord in 1994. At the time, the WTO touted it as "the largest trade negotiation ever, and most probably the largest negotiation of any kind in history."[8]

TRIPS set the global policies for how companies can trade the products of biotech — in particular, commercial transgenic seed and food — across national boundaries. TRIPS remains a sore spot for countries that are concerned about the geopolitics of food and hunger, that biotech industry superpowers like the U.S. will use TRIPS to colonize the market for seed and other vital commodities. One African ambassador, speaking at the infamous WTO meeting in Seattle in 2001, said the TRIPS accord will "create the potential for disastrous conflicts between the technologically advanced and the less technologically advanced countries. It will endanger the traditional rights of farmers and of local communities all over the world."[9]

Why should it be any business of ours, from the perspective of risk, whether farmers are able to save seed? If they sign these technology agreements of their own free will, why is it our concern? Besides, given the rapid clip at which farmers are planting transgenics around the world, the licensing restrictions don't seem to be stopping them, so they must not be too worried about their so-called "traditional rights."

But let's think more broadly. First, keep in mind that farmers may not be getting the quality of information that they need to make good decisions about planting transgenics. We all do our own risk calculations based on the information we have access to. In this case, farmers might be minimizing their concerns about losing control of their seed and their land based on the health and economic benefits they've been promised — benefits that, based on what you've just read, might be considered dubious, or at least not worth the tradeoff.

In addition, in the big picture, saving seed is life insurance, literally. The ability to save seed is arguably the single most important component of food security; it enables an individual, a community or a nation to feed itself. A bag or a bushel of saved seed, particularly for poor farmers, may not be money in the bank, but it may be the only food they can count on for the coming year. And while the problem is more obvious when you look at it through the eyes of poor farmers, it's just as true for those of us who live in the developed world. Surely we should be contemplating the long-term risks as well — risks to farmers, to those of us who eat the food they grow, to individual countries, to the practice of democracy — if some significant percentage of the world's farmers become dependent on having to buy seed each year from a handful of multinational corporations. Surely we should be concerned about how our lives will change if corporations are able to effectively control the food supply.

Again, no official cost-benefit analysis I know of has ever included or addressed these issues. And that's because cost-benefit analysis in the real world is about power. Those who control what goes into the analysis also control what comes out of it. This is a dangerous state of affairs for ordinary people when, in so many countries around the world, industry and government regulators are often on the same side of the negotiating table — as is almost universally true when it comes to commercial biotechnology.

As a result, it seems clear that we can no longer trust cost-benefit analysis, especially as practiced by our regulators today, to honestly and accurately inform us about risk. In the same way that genetic engineering doesn't live up to its warranty of precision and predictability, today's cost-benefit analysis is too vulnerable to bias and manipulation to be a valid measure of the complex social and scientific risks we confront. Even though I suspect many risk analysts will strongly disagree with this perspective, cost benefit's focus on monetizing risk — on defining cost as "amount paid" and benefit as "amount earned" to the exclusion of other important factors — is actually at the core of the controversies around transgenic products.

In theory, using money as the common denominator for cost makes perfect sense. As Warner North and many other risk analysts of his experience and stature have said to me, money is just one yardstick, albeit a broadly accepted one, for value. But in practice — in a market-driven culture and economy, of which America is the global exemplar — that yardstick is more like a hammer.

The downside of monetizing risk and benefit in the face of pervasive scientific uncertainty and such profound social implications turns out in practice to be exactly analogous to the concept of "strongest available science." It can be just as ill-considered to give too much weight to economic and financial implications in a risk calculation as it is to favor an overly mechanized view of biology, and for the same reason. We're back to our preference for "the number" again — overestimating the importance of something just because we can measure it or trade it as a commodity.

We know this intellectually, yet our desire to take advantage of the persuasive power of The Number almost always blinds us to the more subtle, equally important impact of other factors, ones that can't easily be assigned a monetary value.

Given that genetic engineering is "one of those things that human beings have messed around with and arguably might have made worse," as North said to me in a conversation, economics should be only one consideration in a cost-benefit analysis. What's much more important is to include all the ways we can think of that biotechnology can help or harm us, using *all* the resources and *all* the expertise at hand, beyond the strictures of molecular biology or purely monetary representations of risk and benefit.

"Once we've been honest and realistic about the benefits of genomics and agreed upon them," North told me, "then we can look at the dangers involved — and start to direct research toward understanding and solving those problems."[10]

But we are a long way from such a sensible approach. As a result, the dangerous shortcomings of cost-benefit analysis as it's officially practiced today have inspired some scientists to reject it in favor of something wholly different. According to Mary O'Brien, a botanist and environmental scientist based in Oregon, the problem is that cost benefit is almost always used to answer an unnecessarily limited question, namely, she said, "How much of this hazardous substance or activity is safe — or of insignificant, or at least 'acceptable' risk?"

The more appropriate question, according to O'Brien, is, "What is the least hazard that is necessary to solve the problem?" That is, not how *much* is "safe," but how *little* intervention can still accomplish the goal?

O'Brien has worked with many government agencies and other organizations over the past 30 years to help develop policies for managing and sustaining the quality of the environment.

"I have repeatedly witnessed the scenario of someone putting forth a hazardous proposal, analyzing it with complex models and assumptions, and then defending the proposal with those numbers," she said. "We — members of the public, advocates of human health, advocates of wildlife and ecosystem health — are subjected to these technical risk assessments and asked to accept their conclusions of insignificant harm, when we often know there are far wiser alternatives that are not on the table."

For example, she has been involved for several years in the issue of whether the U.S. Army should incinerate the chemical and nerve gas weapons it has stored in northeastern Oregon. In 1996 she urged the Oregon Department of Environmental Quality (DEQ) to consider alternatives to incineration. She had good science-based reasons to be concerned: Incinerating weapons that contain chlorine would inevitably produce and release dioxin, an extremely toxic carcinogen, and other weapons would release toxic heavy metal vapors.[11] Knowing this, the U.S. Congress and the Army had funded a research program to establish what they hoped would be an ideal weapons-destruction technology. Four sound methodologies resulted, at least one of which would be able to destroy the weapons in Oregon without production of dioxin or lofting of heavy metals, while requiring less water than incineration.

Meanwhile, back in Oregon, a risk analysis with "an impressive number" of charts and graphs calculated the risks of incineration, said O'Brien. "Of course the risk analysts estimated the capability of [air] scrubbers to capture most of the dioxin and heavy metals." In fact, they'd modeled and calculated and estimated the potential of every conceivable variable, from nerve gas releases to the deposition of metals and gases in local water sources.

But "at no point was a risk analysis run on any non-incineration technology compared to incineration," she said. "Nor were the two types of technology ever compared for their comparative costs and benefits." And so the Oregon incinerator began burning weapons in 2004, while in Colorado, after comparing incineration and non-incineration technologies, the community and the Army opted to use neutralization.

"So we reject the risk analyses," she said of herself and other environmental scientists. "It isn't that none of us understand models and formulas and the data which are cited. It isn't that none of us understand that we face larger hazards from some other sources. It isn't that we want a risk-free world. It isn't that we're not rational. But many of us know that the hazards foisted on us, or on wildlife or communities or infants or grasslands or water systems, aren't necessary."

"And when the risk analyses don't deal with that fact," she said, "they become tools for bullying."[12]

≡

O'Brien's perspective raises a broader question I used to ask myself when I was writing about computer technology and the "next new thing" would cross my desk. It's a question that's even more critical in the cost-benefit context of genetic engineering, where we are essentially flying blind as we're dabbling in the genome of the planet: Is this innovation (whatever it is) a "solution" in search of a problem? Have we become so enamored with our tools and our technologies — or probably more to the point, so invested in them — that our priorities are all askew?

The perfect example of this is a transgenic crop that gained fame in large part because its purveyors have promised to "give" it to those who "need" it most: the much celebrated, near-mythical "golden rice."

In 2000, Syngenta scientists announced that they had engineered a strain of rice that contained beta-carotene, a compound that the human body can convert into sight-saving vitamin A. The several biotech companies whose patented technologies went into the invention agreed to "donate" licenses to any developing nation to use the seed free of charge as a humanitarian gesture. As a result of this public display of charity, golden rice was immediately heralded by the media as a way to "save a million kids a year"[13] from blindness. The positive response was so swift and overwhelming that one of the Syngenta inventors publicly proclaimed that anyone opposed to golden rice would be responsible for "millions of unnecessary blind children and vitamin-A deficiency deaths."[14]

Nevertheless, golden rice skeptics still came up with some research reports of their own that contradicted the inventors' promises. Infants are the most vulnerable to blindness from vitamin A deficiency; based on industry data, they calculated that mothers would have to eat almost 40 pounds of cooked golden rice per day in order for their breast milk to contain the requisite amount of the vitamin.

Moreover, if golden rice were simply substituted for a daily diet of conventional white rice, they said, a child or adult would receive only 8 percent of the daily vitamin A requirement. And while this modest percentage of the recommended daily allowance could "lift millions of people out of sub-clinical vitamin A deficiency,"[15] as one report put it, the problem golden rice was supposed to address was actually *clinical* vitamin A deficiency — which is what causes blindness. What's more, the human body can convert beta-carotene into vitamin A only if enough fat and protein are also part of the diet. Malnourished people lack fat and protein in their diets by definition.

Faced with such thoughtful challenges to their proposed benefits, even the inventors of golden rice had to concede that the transgenic grain could be helpful only if it were part of a larger complement of foods — again, something of a conundrum when chronic malnutrition is one reason for vitamin A deficiency in the first place.

Syngenta claimed in 2004 that it had developed new strains with much higher levels of beta-carotene. Even so, there are still many unanswered questions about its utility and its safety, from both a health and an environmental perspective.

First, distinct from the need for a varied diet are other unanswered questions in the scientific literature about just how much and how well the various types of beta-carotene in golden rice are converted into vitamin A by the human body. As a result, no one knows how much vitamin A will actually be absorbed by those who eat golden rice.[16] Other reports show that when stored or cooked, golden rice loses its beta-carotene, thus its ability to supply vitamin A. How can this be useful?

Many of the attendant health and environmental risks of transgenic golden rice will sound familiar by now. For one, rice is an avid cross-pollinator, and like with maize, wild and weedy relatives often grow near to domesticated varieties. This raises concerns about the contamination of traditional seed stocks and, of course, about the unwitting creation of yet another transgenic monoculture. Monocultures are bad news even without the pesticide resistance problems that have begun to plague *Bt* and Roundup Ready crops. But a golden rice monoculture could also pose health problems for humans and who knows what other organisms if its transgenes end up being unstable over time.

Already, some of the golden rice plants are behaving in unexpected ways,[17] and unexpected compounds have shown up in the grain.[18] No one knows what other compounds might form, and under what conditions, that could be anti-nutritional, allergenic or even toxic in humans. Scientists don't even understand how golden rice got its color. According to its inventors, the genes in the transgenic cassette should have made the rice red, not yellow.[19]

Because there would be no controls of any sort on the technology in particularly needy countries — i.e., farmers who received the seed free as "humanitarian aid" would not need to go back to Syngenta to buy new and improved seed every year — any unanticipated consequences could hit hardest in the very populations that are the least equipped to respond.

The short history of golden rice is especially peculiar considering the circumstances under which Syngenta announced, in 2004, that it was halting all its European field trials of transgenic plants and seed varieties. (It transferred all its biotech research activities to the U.S.) Its research director cited public resistance, a hostile regulatory environment and the lack of "market opportunities" as the reasons for abandoning ship. But in the same news article, he was quoted saying that Syngenta had "conducted many genetic engineering experiments for seed materials and plant protection and [these experiments] have often failed." In fact, he said they'd often found conventional methods to be more effective than biotechnology.[20]

The golden rice seed cost $2.6 million to develop, and it will cost a further $10 million to adapt it to local conditions.[21] While vitamin A deficiency is still a global health concern, a 2004 report by experts in the area of micronutrient deficiencies said that "very significant progress has been made over the last 15 years" to combat it.[22] Already 43 countries now have formal supplementation programs that reach at least two-thirds of all young children in these countries. And according to a Greenpeace report on the issue, 10 countries have virtually eliminated vitamin A deficiency.[23] Given these challenges to its promised benefits, golden rice looks suspiciously like a solution in search of a problem.

The world of commercial biotechnology is rife with examples like golden rice. Examples of failed transgenic experiments from the world of animal engineering are particularly poignant in this context. Data from the now-defunct PPL Therapeutics, Inc., the Scottish biotechnology company that specialized in designing transgenic animals (including Dolly, the sheep), showed that only 134 calves were produced out of more than 25,000 fertilized eggs that were microinjected with transgenes. Of those, only nine of them turned out to be transgenic.[24] (Microinjection is an alternative to plasmids as a technique for flooding a cell with transgenes: it involves directly inserting DNA into the cell with a very fine needle.) Data from a pig trial revealed that after injecting a specific transgene into 7,000 ova, less than 1 percent of the newborn piglets carried it.[25]

In 2002, the National Academies published a study on another form of genetic engineering, human reproductive cloning, and it contains what today may still be the only comprehensive public data that has been collected on transgenic animals. It reviewed all the available data on the outcome of cloning mammals, including sheep, cattle, goats, mice and monkeys. Of a total of more than 12,000 transgenic embryos, only 207 of them, less than 2 percent, resulted in live births. Transgenic animals that didn't turn out as designed generally didn't live long and are much more controversial than transgenic plants.[26] Those that were carried to term and lived had many physical and developmental defects, including dysfunctional immune systems; lung, kidney and circulation problems; and liver, joint and brain defects. Poor famous Dolly died young, with premature arthritis and lung disease for all her trouble.

Most unfortunately from the perspective of risk and uncertainty, "these abnormalities have not always been studied in detail, possibly because animal cloning has been done for commercial purposes and there is less interest in the failures than in the successes," according to the Academies' cloning study.[27] Based on present scientific data, it concluded that "success is not a reproducible phenomenon, and ... the precise molecular mechanisms responsible for the high failure rate are almost entirely unknown."[28]

The scientific concerns that were raised in Chapter 4 about the biological confinement of transgenics take on even more depth in the context of cost and benefit. Isn't there a better way to "grow" more salmon than changing their genetic structure and then sterilizing them so they don't take over the ecosystem? The idea that we could ever make biological confinement worth the risk both strains credibility and offends common sense.

First of all, we've already accepted what should be a wholly unacceptable condition — that we have created something that we cannot physically confine and that will wreak irrevocable havoc when it escapes. Then we attempt to "biologically" confine the first mistaken creation, by mucking around more in the same genetic material we couldn't predictably control in the first place. How precisely do we plan to avoid the inevitability of this new intervention creating yet another ecological fire we'll have to put out, with yet another "solution" that's certain to create yet another problem?

Or have we lost sight of the problem entirely?

It's hard not to ask that question once you've looked hard at the reality of transgenic products that are already unstoppably circulating among us — although so far, anyone who has tried to mount an inquiry into the topic has been felled by the flaming twin arrows of selective scientific evidence and American economic imperative. Even as eloquently and reasonably as Norman Ellstrand presented his doomsday scenario about transgenic pharm crops, he still stopped short of saying, "These are a bad idea."

If in fact there is a problem that can be solved by pharming — and I'd require some strongly moderated discussion about that before I'd agree to the premise in the first place — why isn't industry obliged to investigate other, less risky alternatives first? Why would anyone spend more than $12 million to "cure" vitamin A deficiency with a transgenic crop that could contaminate one of the world's staple grains when there are other, safer methods that are already working? And why would any government allow even one of those golden rice seeds to be planted in its soil until the benefit could be *proven* worthy of the risk?

"My singular objection to all the investment and emphasis on transgenics is what it draws attention away from," said Kim Waddell, who directed the bioconfinement study. "No research funding is being invested in other kinds of tools or solutions. If we invested just 10 percent of the USDA's research budget into other areas, we'd be tripping over ourselves in discoveries and revelations. But with all the money that goes into monocultures and transgenics, we've gone too far down the road for industry to adopt alternative approaches — even if a different approach might do a better job."[29]

CHAPTER 10.

OUR APPOINTED ARBITERS OF RISK

All things considered, we're more than just "too far down the road" with transgenic technologies. I'm not sure we even know what road we're on; we're driving too fast to read the signs.

It didn't have to be this way. The same problems that have kept us from a defensible risk analysis on the products of genetic engineering are the same ones that have long plagued the assessment of all risky technological innovations: not having enough data to analyze, a shifting landscape of scientific knowledge and an ongoing need to balance risk and benefit in a way that rewards innovation without sacrificing public safety. These problems were long ago documented by the experts who study the risks of science and technology. In fact, each of the U.S. agencies that now regulate transgenics has, at various times during the past 20 years, commissioned these experts, from a wide variety of disciplines, to help them figure out how to improve their methods.

Four studies and many spectacularly practical recommendations later, there have been no obvious changes in the regulation of technological risks. Certainly none of the methods these studies have recommended have been applied to official regulatory assessments of the risks of transgenics.

Why not? There isn't a simple answer to that question. But one part of the explanation has to do with a critical variable that I've addressed only indirectly so far. The crux of the problem, practically speaking, is the role that technical experts play in official risk assessments.

Our sophisticated technological societies have become almost wholly reliant on expert specialists to set the agenda for scientific progress. These experts are confident that they're entitled to the authority they've been granted by society to determine "truth" within their areas of research and development, including what technologies in those areas are risky and how those risks should be addressed. Like anyone who has amassed power, experts tend to protect their position if they believe their authority is being threatened. But there are many reasons to challenge the dominant "trust us, we're experts!" paradigm that now rules the practice of risk assessment — not just for genetic engineering, but for all areas of science and technology.

≡

Within a few months of the historic 1975 Asilomar meeting on the risks of recombinant DNA (discussed in Chapter 2), two of today's best-known risk experts — Baruch Fischhoff, the Carnegie Mellon risk professor introduced earlier, and his colleague, psychologist Paul Slovic, president of Decision Research in Oregon — also found themselves at Asilomar. But they were attending a very different confab from the one organized by the DNA experts. The subject of the latter Asilomar, which was sponsored by an engineering foundation, was an exploration of how best to use cost-benefit methods for analyzing "low probability, high consequence" events — disasters of the very same type that the DNA experts had been trying to foresee if one of their recombinant experiments were to go awry.

At the engineers' Asilomar, the cataclysm du jour was liquefied natural gas, which to this day continues to be evaluated for safety as a potential alternative to fossil fuel. As its name implies, liquefied natural gas has been changed from its gaseous state to liquid in order to make it easier to transport and use. The process involves chilling or pressurizing the gas, then vacuum-sealing the liquefied product into a container. The question on the table then, as it is today, was how the technology might behave at scale. It was easy enough to experiment with different ways to make the gas explode (or not) while it was sealed in a little lab-sized container. But what method could be used to find out whether it was safe to ship and store commercial-sized volumes of such a volatile substance — like tankers full of it, for example — and under what conditions?

As you might expect, the engineers presenting data at the conference had approached the problem from every technical angle they could think of, but for Fischhoff they kept missing the mark. "Here were all these analyses on the safety of the technology, and they all looked good," he said. "Except I felt uncomfortable about the assumptions they were making about people's behavior. They were assuming perfect maintenance of the equipment and the technology — that nobody would do a bad job, or forget to replace a part."

Then, said Fischhoff, one of the participants would say something like, "'Well, we really don't know how metal will perform under these circumstances,'" he recalled. "And then someone else would say, 'We really don't know how large volumes of liquid natural gas will behave when released.' And, 'We don't really know what would happen if the gas was accidentally released and floated toward Manhattan. Would it form a cloud? Would it blow up if somebody lit a cigarette?'"

As a psychologist, said Fischhoff, "I thought they made unrealistic assumptions about people's behavior. And from a mathematics perspective, I saw that even their technical analyses were rife with judgment." Once those kinds of expert judgments end up in a cost-benefit equation, they look like Paul Thurman's probability-spouting economist from Chapter 5 — calling out a quantitative, "30 percent" chance of a recession as though the forecast were based on data, even though it's not. At the time of this meeting, the use of cost-benefit analysis by government regulators was on the rise; Fischhoff thought it was important that both the risk analysts and the "non-expert consumers" of their analyses (that is, regulators and politicians and their various constituents) understand the kinds of errors that the method was prone to produce — all of which were about mistaking human judgments for fact.

In fact, Fischhoff's studies were among the earliest to show that subjective judgments made by experts have just as much effect on the accurate assessment and analysis of risk as hard scientific data.[1] "The evolving insight for me," he said, "was that the real interesting problems are about methods — how you actually do the analysis, when there's so much uncertainty and judgment on the part of the scientists who are dealing with these types of problems."[2]

Yet expert declarations are often the factor that determines what technologies or products are officially deemed safe — even in disciplines that are relatively new and where the collective knowledge is changing all the time, as is the case with genetics and biology in general. In fact, if you *aren't* an expert (by their definition), your perspective on any given risk is easily and almost always dismissed. Scientific judgments may be "more" defensible because they're based in technical knowledge, but as both Thurman and Fischhoff have observed, experts are often just taking their best guess. This is especially true of decisions about potentially cataclysmic risk events that must be made with no historical data at all, like what kinds of things could happen to a tanker of liquid natural gas or, for that matter, what it would take for a population of transgenes to trigger a threshold effect in some population or ecosystem.

The systematic approach of the scientific method — the testing and retesting of hypotheses, the practice of having disinterested reviewers check each other's work — is supposed to neutralize some of that bias. And sometimes it does. But for all the important discoveries it has enabled, the scientific method also breeds a more tenacious sort of expert bias, a consequence of the natural tendency for the tribes of academic disciplines to stick with what they've been trained to know and do. The intensive training that a "disciple" receives in his or her discipline's theories, data and methods is designed to protect the field from invasion by bad actors who just want to make a name for themselves and from sloppy researchers whose experimental results can't be reproduced.[3] The practical result of this training is that the very act of becoming an expert is hidebound with bias.

Expert bias explains why specialists in any field very often have a hard time accepting as valid any knowledge or data that wasn't gathered by methods they trust or that doesn't dovetail with their disciplines' tenets.

≡

Blindness to value judgments about risk factors beyond an expert's purview are obvious hazards. But there are other, less overt biases at work in many of the risk analyses of genetic engineering. These kinds of conflicts, often purposely hidden from public view, are made by regulators and academic scientists under the influence of financial interests, divided loyalties or other undisclosed motives. Experts under the influence of these kinds of conflicts can produce flawed science, and as a result, their analyses and decisions about risky products and processes can be irrelevant — or worse.

"Where the science comes from for risk regulations gets back to a broader set of issues about what academic scientists are up to these days," said Michael Rodemeyer, the former director (and now a senior advisor) of the Pew Initiative on Food and Biotechnology at the University of Richmond. "And to a certain extent, it raises questions about the academic researcher, particularly in molecular biology, who's allowing himself or herself to be captured by these industrial issues."[4]

What academic researchers are mostly up to these days resides in a murky area between applied and basic research. Applied research uses a scientific discovery as the springboard for a commercial product or market opportunity. It's designed for the purpose of producing results that can be quickly applied to real-world situations. Applied research stands in contrast to "basic" research, which is harder to commercialize and is focused on fundamental scientific discoveries. But in both biotech and academia, particularly in the computer and life sciences departments, basic research is increasingly under pressure to produce results that yield patents or technology licenses. This trove of intellectual property can greatly enrich university coffers, by requiring licensing fees from other researchers as well as when professors start companies by licensing back the technologies they developed with taxpayer money.

This type of research generally falls into the category of the "industrial gene," a concept that is the cornerstone of the biotechnology industry. The industrial gene can be owned, in legal terms — i.e., it can be patented and protected by intellectual property laws — because it can be unambiguously described: a sequence of DNA with a beginning, a middle and an end falls rather neatly into that category. These laws also apply much more broadly, to virtually all of the rest of the products and subproducts of genetic engineering, including "re-engineered" molecules and organisms, as well as processes like recombinant DNA itself, the methods used to amplify DNA or find new genes, and the lab techniques that initiate gene silencing.

In fact, an October 2005 article in *Science* noted that fully 20 percent of the genes in the human genome have been patented, and some of the genes have been patented as many as 20 times each as they've been "improved" by genetic engineering.[5] No matter that the much-touted linkage between individual genes and disease is largely a statistical error; genes associated with specific health issues and disease are still popular targets for patents.

Taxpayer-funded research has become a proxy for personal investment and private financial gain by researchers and the universities that employ them.[6] As of February 2004, 10 of the top 25 holders of U.S. biotech patents were universities. The University of California held the top slot with nearly 900 patents, even more than the U.S. government, which came in second.[7] A 2000 investment survey showed that former and current faculty members from just five California universities had founded more than 300 biotechnology companies based on their academic research.

Whether you think that university patents are a spur to innovation or a perversion of the higher societal calling that public universities once represented, the practice has serious consequences from the perspective of risk. That's because as a result of this commercial focus, much of the taxpayer-funded research that used to be widely published — thus open to challenge and debate and further discovery by other scientists — has become increasingly shielded from challenge and debate in the form of patents and "confidential business information." Universities that once refused to do research for any price if it couldn't be published in the open scientific literature are increasingly willing to sell their work outright to the highest corporate bidder.

One of the most contentious examples was a deal signed in November 1998 between the Swiss pharmaceutical giant Novartis AG and the University of California, Berkeley. Under the terms of the agreement, Novartis would give the public university $25 million over five years to fund research in the Department of Plant and Microbial Biology, one of four departments within the College of Natural Resources. In exchange for the $25 million, Berkeley granted Novartis first rights to negotiate licenses on roughly a third of the department's discoveries — including the results of research funded by state and federal sources.[8]

The agreement caused a deep rift within the department's faculty as well as within the entire university and beyond, but it was pushed forward and approved nonetheless. A committee of reviewers from the University of Michigan who were brought in to do a postmortem several years later declared that the cash didn't compensate for the broken trust between colleagues.[9]

But the Novartis deal is only one example in which the intellectual competition that used to inspire scientific discovery has been replaced by the desire to squirrel away taxpayer-funded discoveries and data for later commercial use. Because of the patenting frenzy, researchers in the near future may find it costs them considerably to gain access to gene-oriented technologies. Universities are so focused on generating profits that they often demand that even researchers in the same department or on the same campus apply for licenses, just to get access to the basic scientific knowledge generated by their colleagues. According to Jennifer Washburn, an expert in university-industry conflicts, in a 1997 survey of more than 2,000 members of life sciences departments, 34 percent of those who responded claimed they had been denied access to research results or products generated by their academic peers.

This commercial chilling of scientific inquiry has spread beyond the department and into the pages of the scientific journals that were once seen as lifelines to the latest discoveries and knowledge. A 2004 editorial in the journal *Nature Genetics* beseeched academic and corporate researchers to start releasing their proprietary data to reviewers again, so they might receive the kind of scrutiny that is required for the conduct of good, credible science. "We note that there is now no clear line between academic and corporate research," said the editorial, which was titled "'Good citizenship' or good business?"[10]

Asking researchers to consider choosing professional camaraderie over cash for the sake of science is one way to frame the question. But from the perspective of risk, the fact that it's a choice at all is troubling.

The experts on the scientific staff of government agencies don't seem to be any less susceptible than academics to the lure of cash. In late 2003, the *Los Angeles Times* published an investigative report showing that several ranking officials at the once-unassailable U.S. National Institutes of Health were receiving hundreds of thousands of dollars each in consulting fees, paychecks and even stock options from drug and biotech companies, in addition to their government salaries, which are paid with taxpayer dollars. More than a third of NIH's scientists had profited from the deals.[11]

When NIH leadership responded by announcing it would prepare a conflict-of-interest policy, the researchers responded in kind, saying, "Fine, then we'll leave" for the more welcoming shores of academia, where making money on the side from industry is widely accepted and acceptable.[12] To which I think the most appropriate response is: "Good riddance."

≡

With so many academic biologists openly co-opted by commercial interests, are their consultations or pronouncements about the risks of their technological innovations really disinterested, or are they influenced by their stock portfolios and consulting checks? It's a reasonable question, but the issue raises two other important points about allegiance and priority that wrongly get mashed together in conversations about expert advice.

For starters, companies have a very different role in risk assessment than do either academics or government regulators, or at least they should. The risk to a company selling transgenic products is fundamentally different from the risks to the ultimate user of those products — the person, you or me, who takes the drugs, eats the food, breathes the pollen, drinks the water, etc. If a company's products have been approved by the proper regulatory authorities, it shouldn't have to concern itself with safety.

As Monsanto's public relations chief memorably said to *The New York Times Magazine* in 1998, "Monsanto should not have to vouch for the safety of biotech food. Our interest is in selling as much of it as possible. Assuring its safety is the FDA's job."[13]

He's absolutely right. That's the proper fiduciary attitude for a corporate spokesman. But since we're the ones who are presently eating the results of the FDA's substantial equivalence policy, let's disentangle his first point about Monsanto's priority — selling as much biotech food as possible — from his second point, about the FDA assuring its safety.

What Monsanto's PR chief didn't acknowledge was that most of the opposition to transgenics overall would be neutralized overnight if we could actually trust that our government is doing a proper job of assuring *our* safety — that there are no connections between the industry and government regulators that might compromise or bias the kinds of risk and safety assessments that governments make on our behalf. But the facts as they stand today don't inspire that kind of trust. In fact, there are so many deep connections between biotech companies and the U.S. government that one nonprofit group compiled a list that it calls "Revolving Door," a database of people who have worked both in and for regulatory agencies and the biotech industry.

The Edmonds Institute began compiling its database in the late 1990s when "we began to wonder who exactly was negotiating for the U.S. and what were their goals," said Beth Burrows, the Institute's president.

Its first Revolving Door publication, distributed to negotiators who were developing an international biosafety protocol, announced the job change of the USDA's Val Giddings. In 1997, Giddings left a position as chief for science and policy coordination in a biotechnology division within the USDA; the following week he joined the Biotechnology Industry Organization as vice president for food and agriculture. BIO is the main lobbying group for the biotechnology industry.

Giddings is only one of many who have made such changes to and from government and industry positions. Edmonds' Revolving Door tracks more than 100 people who have shuttled back and forth between industry and government agencies at various levels of responsibility. Another notable example is Michael Taylor, who was instrumental in helping develop the substantial equivalence doctrine. Taylor left his position as staff lawyer at the FDA in 1981 to become a partner at King & Spalding LLP, a law firm that specializes in counseling corporate clients "on a full array of FDA and related matters"[14] involving biotechnology and other FDA-regulated industries. From there he joined Monsanto as vice president for public policy. Taylor left Monsanto in 1991, returned to the FDA as deputy commissioner for policy, and stayed until 1994. During his tenure the substantial equivalence policy became law; he is often acknowledged as its primary author. (Taylor is now director of the Center for Risk Management at Resources For the Future, a nonprofit policy institute in Washington D.C.)

Monsanto also looms large in the resume of Linda Fisher, who has held several senior positions at the EPA, including deputy administrator and chief of staff. She started at the agency in 1985, worked for Monsanto as vice president of government affairs from 1995 to 2000, then rejoined the EPA as deputy administrator in 2001. After she resigned from the EPA in 2003, she joined DuPont.

Ann Veneman, who resigned as secretary of the USDA in 2004, has long interspersed her high-level regulatory positions with jobs or board positions that serve the international interests of U.S. agriculture and biotech companies. This included a stint as legal counsel with a firm specializing in representing agribusiness giants and biotech corporations. She's also an emeritus member of the International Policy Council on Agriculture, Food and Trade, a trade group funded by agribusiness giants and lobbyists including Archer Daniels Midland Company, Grocery Manufacturers of America, International Fertilizer Industry Association, Monsanto, Nestlé Corp., Syngenta and Unilever N.V.[15]

Shuttling back and forth between industry jobs and the agencies that regulate them is perfectly legal. It has been going on for decades all around the world in commercial areas other than biotech. But when the technologies are risky by nature and the stakes are high on both sides, the influence such people can wield over regulations takes on heightened importance. As Burrows of the Edmonds Institute said it, "Sometimes people question exactly when the interests of their public servants stopped being the public interest." And even a superficial look at some of the U.S. regulations for transgenics makes it pretty clear what side of the public-private divide the U.S. government is on.

For starters, "the agencies don't provide information about specific risk assessment methodologies [about transgenics], or any good public explanation of what their methods of analysis actually are," said Rodemeyer of Pew. In 2001, his organization published an exhaustive *Guide to U.S. Regulation of Genetically Modified Food and Agricultural Biotechnology Products.*

"You can look at the consultation letters that the FDA publishes and can see the process by which they analyze the data. It's certainly some kind of structured analysis, but it's all within the confines of what's 'generally recognized as safe' and substantial equivalence," Rodemeyer said. Much the same is true for both the EPA and the USDA. "The real question is exactly where do the agencies get their science from, even the science to frame the questions they should be asking."[16] And not just the science, he said, but the data to support it.

Okay, I'll bite: where *do* the agencies get the scientific data they use for their decisions about transgenics?

When they bother to acquire any data at all, the short answer to the question is, "From the biotech companies themselves."

≡

Not one of the U.S. agencies that regulate transgenics — the EPA, FDA, and USDA — is required to generate its own data for risk analysis, and rarely do any of them conduct original research. Every agency's safety reports are prepared by reviewing the product data that the company provides. Independent testing on transgenic products is not part of the regulatory process, nor is independent confirmation of the accuracy or the quality of the corporate data that's submitted to them. Most of the time they wouldn't be able to independently confirm the data even if they wanted to, as the data contains "confidential business information" that the developers won't release unless compelled by law.

For some products, the biotech company has to provide data that proves its products fit within regulatory guidelines. But for others, the onus is on the agency itself; it must prove that a company's product *doesn't* comply. And while regulators and biotech companies are very solemn about their duty to protect public safety and public interest and about how their assessments are based on "rigorous science" and so forth, a closer look does not provide much comfort that the risk analyses of transgenics is taken seriously.

For example, a 2003 report from the Center for Science in the Public Interest (CSPI) used information from the FDA's own Web site, in conjunction with a Freedom of Information Act request, to collect all the publicly available information from 14 of the 53 voluntary product "consultations," as they are called, on transgenic crops that have been submitted to the agency.

The CSPI report evaluated several consultations, from different developers of *Bt* corn, transgenic tomatoes and cantaloupes, and others. Its assessment showed that when the FDA asked for more information to complete a safety assessment, three out of six of the developers didn't bother to respond. They really didn't need to. Because the notification was voluntary, the FDA had little choice but to complete its evaluation without the safety information. Furthermore, three out of 14 of the data summaries submitted by the developers contained obvious errors that the FDA didn't even flag during its review process.[17]

The FDA is the birthplace of substantial equivalence, granted, so its laissez-faire attitude is to be expected to some degree. But what about the Environmental Protection Agency, whose mission to protect is built into its name?

The "Plant Incorporated Protectants" page of the EPA's Web site states that its scientists have "assessed" and "evaluated" a wide variety of potential effects that could be associated with transgenic plants. They've tested for "acute reactions, such as toxicity, allergenicity," as well as for other irritants. They've assessed the transgenics under their purview for "long-term effects including cancer, birth defects, reproductive and neurological system" problems. And it says they've taken into account all the combined sources of exposure to the transgenics they regulate, including drinking water. Based on their "reviews of the scientific studies" and often peer review by another federal scientific advisory panel, they've determined that the transgenic products they've approved "would not pose unreasonable risk."[18]

But according to a former EPA scientist, the EPA does no routine long-term testing for transgenic crops, not even the tests that are required for chemical pesticides that are applied rather than incorporated into a plant's genes. It can apparently make these claims by using a little scientific sleight of hand. In its fact sheets on individual crops, the EPA often cites a paper written by several of its own scientists: Administering high-dose tests, it states, is adequate to account for possible long-term effects of transgenic proteins.[19] A high-dose test is supposed to mimic the effect of a living organism ingesting the protein over time.

Whether or not high-dose tests are in fact an adequate substitute for long-term testing is subject to a debate that I won't engage in here. The bigger question is whether these "toxicological considerations," as the paper calls them, should be considered the most important environmental risk factor for transgenics. Is it an example of "what gets measured is what matters?"

Traditional measurements of chemical effects — acceptable levels of toxicity, parts-per-million and the like — wholly overlook environmental risk factors that are unique to transgenics and genetic engineering. For example, testing a transgenic protein, at a high dose or otherwise, tells us nothing about what happens over time as the millions of transgenic plants in the field reproduce, spread and evolve. What genetic changes do they undergo under normal circumstances, and what are the effects on the health of the plants as well as on the proteins they produce? What happens if they are subjected to unusual environmental stress, or when disease strikes? Can biotech developers predict whether the genomes of their transgenic products will remain stable in the field?

Unpredictability is risky by definition, but it's a serious hazard in biological systems. That's because when a DNA sequence of any origin — transgene or native — fixes in the wrong place in the genome, breaks into fragments, or somehow moves from its original location, it can disrupt the expression of neighboring genes, or change the way it is itself expressed, in unpredictable ways. These so-called position effects can include the unexpected silencing of native genes; recall the case in human gene therapy, where the therapeutic gene landed in the wrong place and triggered a fatal cancer.

Researchers have already observed these "transgene rearrangements" in the lab. Some have noted them in transgenic oat plants.[20] Others have recorded mutant offspring in transgenic lettuce, rice, maize and barley during lab experiments where they were inserting transgenes into several locations on the plant's genome.[21] Even if they don't produce obvious mutations, unstable transgenes could go unnoticed, causing long-term damage to plant populations — silencing genes, expressing new proteins, and otherwise raising entirely new questions about food security and plant and animal health that no high-dose test can answer.

But it hardly matters today. The EPA states plainly that it regulates only the protein and its genetic material, not the plant itself. The potential effects of unstable transgenes aren't officially considered at all.

Finally, what about the USDA, the agency that is responsible for the safety of meat, poultry and eggs? The agency that is the steward of 192 million acres of America's national forests and rangelands? What kind of science does it use to regulate the safety of the transgenic products under its purview?

As it turns out, the USDA is in a class of its own when it comes to biotech regulation. The agency is so convinced of the safety of genetic engineering that once a developer has assured the agency that a transgenic plant will not become a "plant pest," the USDA not only deregulates it, but relinquishes all regulatory authority over it forevermore. At that point, it can't even monitor the crops for unanticipated problems. In fact, Monsanto's technology license for its transgenic seed gives it more authority to monitor the fields and crops of farmers than the USDA has.

But even unabashed advocacy doesn't seem to account for the laxity of some of the USDA's decisions about transgenics. In 2003, the USDA granted a preliminary version of the "nonregulated status," as it's called, for an herbicide-resistant variety of Roundup Ready creeping bentgrass, the one developed by

Monsanto and Scotts mentioned in Chapters 3 and 4. They had done so even though bentgrass has already well proven its predilection to breed with noxious weeds. And while transgenic bentgrass may not *technically* be a plant pest on its own, with its built-in resistance to a weed killer and its propensity to pollinate noxious weeds — and possibly pass along that resistance to its hybrid progeny — it's hard to imagine anything safe about this product.

An outcry from farmers as well as concerns from the EPA caused Monsanto and Scott to shelve their plans to bring it to market, but the USDA had already granted its preliminary approval. You don't have to look far to find the reason: USDA scientists aren't just regulating biotech developers, they *are* biotech developers. The USDA is actually named on the patent of the "terminator" seed sterilization technology that's proposed as a method for "biocontainment" of transgenics. But that's only one of its most recent contributions. For more than 100 years, the agency has been importing new strains of fruit, inventing and deploying new herbicides and insecticides, finding replacements for harmful insecticides, inventing or importing new higher-yield or disease-resistant plant and animal breeds, and making key discoveries and inventions in just about every area of agriculture. The agency often partners with commercial interests to develop its discoveries. In fact, in 2003 it spent nearly $180 million on biotechnology research and development.

Yet less than 2 percent of the agency's budget — $3.4 million — was routed to the agency's Biotechnology Risk Assessment Grants Program to study the environmental and food-safety impacts of the products that it and others develop.[22]

"If you look at USDA as an early proponent of biotechnology, you can maybe begin to understand the ambivalence there has been about saying 'Wait a minute, maybe it's not all so good,'" said one USDA insider. "We're working on the future of agriculture. Even if another agency were to say, 'Let's go slow on this,' the scientists here say, 'Why go slow? These are all good things. We developed them to be good and they are good.'"

The sentiment may sound familiar. Remember Todd La Porte's comment early on, about the nuclear engineers who didn't want to believe that their activities, which were done for all the right reasons, could actually cause harm?

"To think 'other people might suffer as a result of my actions' is not part of their world, or it gets pushed away in the drive to deploy," he said of nuclear engineers, although it can apply to anyone whose job is to create technology from scientific discoveries. "But what are the consequences if it turns out that all the things you believed in are wrong? Most technical people can't talk about this — what they do is theology to them, not science."[23]

I resurface this now because it's a key point for how and why experts themselves have driven the conversation about the risks of transgenic organisms so far out of whack. Keep in mind what has been learned about science and judgment, and it becomes obvious that La Porte's nuclear engineers, aside from the inevitable exceptions to the rule, weren't lying about the safety of nuclear power, not even to themselves. They weren't championing new power plants because they wanted to keep their jobs more than they wanted their friends and neighbors and users of nuclear power around the world to be safe. The expert judgment of the very best of their tribe had rendered them absolutely certain about how a nuclear power plant worked, and that's what they believed and acted upon.

They were so certain, in fact, that they were also convinced that the people who disagreed with them were the ignorant ones. This same certitude blinded them to critical uncertainties, like human factors, that to them seemed wholly irrelevant to the safety question.

The official stance of "experts" in the context of genetic engineering seems very much the same as in those pre-meltdown days. The biotech triumvirate composed of academia, industry and government agencies is certain that it is right about the safety of genetic engineering. Its disciples believe that they know more than enough about the science to say it's safe and that those who oppose it either are ignorant or are competitors with other technologies (such as organic farming).

But why should the evidence of opponents be considered any less credible or more conflicted than that of the corporate researchers themselves? Corporations have millions of investment dollars at stake (at least) if their transgenic products never come to market or fail or are rejected by the public once they get there. This may have been the brand of desperation that was behind the behavior of Monsanto executives in Indonesia, who in 2005 were caught having bribed at least 140 government officials — including a $50,000 cash gift to its Environment Minister — to avoid (unsuccessfully) having to provide an environmental assessment for its *Bt* cotton.[24]

Recall, too, the biotech executive who had the "data" for Waddell's proposed study on transgenic benefits, the "director of regulatory science" for his company. Did his title strike you as peculiar? Did you wonder what might make "regulatory science" different from regular old science? Because products are regulated based on risk, one could reasonably question whether regulatory science is actually about how best to perform probative risk analyses — i.e., development of the science to prove that products are "safe enough" to pass muster, in order to move them more quickly to market — as opposed to fully exploring the products' risk profiles.

This isn't wild speculation or aspersion. Experts who have studied how formal risk analysis is employed inside of organizations have shown that it often isn't used as we would expect; that is, to establish quantitative estimates of risk. Instead, risk analyses are often afterthoughts to political and economic considerations, used to rationalize management decisions that have already been made.[25]

≡

With these examples in mind, how has it come to be so consistently true that virtually any scientist or organization that challenges the tenets or products of the biotech industry is subjected to greater public scrutiny than the results of corporate researchers? Corporations rarely talk about the tests they conduct on the products they invent, let alone allow those tests to be independently reviewed and published, or otherwise reveal their results. That almost never happens, unless the results are positive, or they are compelled to do so by some higher authority, whether a regulatory agency or a court of law. Witness the unprecedented defensive action taken in early 2005 by 20 of the biggest chemical companies in the United States. In a move that stunned the academic publishing community, attorneys for the Dow Chemical Company, Monsanto, Goodrich Corp., Goodyear Tire & Rubber Company, Union Carbide Corp. and other chemical giants subpoenaed and deposed five academics for nothing more than reviewing a book and recommending that it be published.

The book, from University of California Press, was written by two highly regarded New York history professors, one from The City University of New York and the other from Columbia University. Its topic was corporate cover-ups of industrial pollution.[26]

Such intimidation tactics and their potential to stop scientists from researching controversial products or behaviors are deeply troubling in the context of risk and public trust of companies that sell products based on scientific knowledge. Very few beyond a relatively narrow band of the scientific community really understand how the products of industrial biotechnology work. The rest of us are fully reliant on the quality of the information that we get from technical experts to make our decisions about whether or not what they're selling us is safe.

Whether or not we *should* rely upon this narrow band of experts is what is at question. A University of California, Berkeley, library exhibit on the 25th anniversary of the founding of the biotech industry displayed two newspaper clippings describing the "gold mine" that biotech patents represent to research universities. "In contrast to 25 years ago ... it is the rare faculty member in biomolecular science who does not now have a relationship of some kind with industry," the annotation read.[27] Even more chilling, according to a 2003 article published in *Nature*

Biotechnology, are trends that "indicate that [university scientists] with corporate sponsorship are more likely to produce favorable findings and to withhold data from the scientific community to protect proprietary interests" than those who don't accept corporate funding.[28] In 2004, a Harris Poll survey showed that scandals about undisclosed financial ties between academic researchers and drug companies are pulling down the reputations of U.S. medical bodies and academic journals with it.[29]

As the sociologist Ulrich Beck wrote in his landmark essay, *Risk Society*, "We no longer pick the experts — instead, the experts choose their victims," based on the information they give us.[30] What they know literally determines what we know. This fact is particularly poignant in the realm of science and technological risk, where the way in which we define expertise and how society rewards its experts determines how those experts define and assess risk for all of us.

Thus scientific expertise — whose we use, and why — is a critical component in the fair and accurate assessment of modern-day technological risk. Yet the biotech experts have so armored themselves and their institutions with scientific explanations that there's no way and nowhere to have an official dialogue about uncertainty and the limitations of their knowledge, including what other kinds of information — such as what you've read in these pages — might be worthy of consideration. In fact, even when it comes to legal challenges to their expertise, the deck is stacked in the experts' favor.

≡

According to Wendy Wagner, a law professor and expert in environmental law at the University of Texas, government regulators use science to mask the policy judgments they've embedded in their risk assessments. Presenting the public as well as the courts with scientific explanations that bamboozle the nontechnical reader and don't explain their assumptions is particularly appealing when those policies are likely to be controversial.

We don't need to look beyond the archetypal substantial equivalence policy for an example. The scientific explanation given by the FDA was that transgenic food was deemed by the tenets of nutritional science to be equivalent to the food we were already eating. This explanation purposely left out the critical assumption made by the agency: that the process of genetic engineering had no bearing on the safety of its products.

Unfortunately, the U.S. courts don't seem to be a very useful place to challenge the expert skew in science-based regulations. According to Wagner, the courts that review regulations tend instead to *reward* regulators for exaggerating the role of science. When an agency claims its risk assessment is, in Wagner's words, "at the frontiers of scientific inquiry" and it doesn't acknowledge or reveal the policy judgments behind it, the courts defer to the agency. "They back off and don't engage in the technical issues. But if the agency starts talking 'values' or 'policy' or even 'economic implications,'" she said, "the court will become more engaged in reviewing the reasonableness of the agency's judgment."[31]

Wagner is not alone in that observation, certainly not among policy experts or environmental lawyers who have hit the wall trying to challenge regulators. Of course, it's not surprising that legal experts like judges would be more willing to engage in deep, meaningful combat about legal issues than they would about technological ones they don't understand. It's human nature. But the legal system's aversion to burrowing into the substance of science and technology is a situation that the biotech triumvirate has learned to play to its great advantage when it comes to the legislation and regulation of risk.

Some policy experts, for example, have noticed that since the late 1970s, U.S. Supreme Court decisions have been reluctant to second-guess the judgment of agencies on issues of science or technology regulations.[32] So as a hedge, the agencies pack official documents with technical data that may or may not have anything to do with the policy at hand, or write what one policy analyst called "'telephone book' statements bloated with too many issues and details undifferentiated in importance."[33,34]

I can personally vouch for this: I looked through one biotech company's application documents for the approval of a transgenic crop seed, which included hundreds of pages of technical data, very little of it useful for addressing the kinds of issues we've talked about here. And as a result of the Supreme Court's reticence to challenge the agencies, the lower courts end up less likely to question the assumptions and judgments behind science-based regulations, because by precedent any decision they make would probably get overturned.[35]

Even when the courts are forced to engage with companies or regulators about technical issues, the legal precedent has essentially the same inquiry-chilling function that we've seen elsewhere: the courts will go along with whatever is considered to be the strongest available science. The Supreme Court case that set that precedent,[36] known as *Daubert v. Merrill Dow Pharmaceuticals*, started as a well-intentioned attempt to keep "junk science" from being admissible evidence in the courtroom. But since then, it has been widely used to keep juries from hearing scientific evidence in product liability and personal injury cases.

In its decision, the Supreme Court came up with a working definition of "valid" science and expertise to help the courts decide what kinds of scientific evidence were admissible. Scientific knowledge, said the court, has the following characteristics: It uses a reliable methodology that yields sufficient data; it's testable with a low rate of error; it has been published in peer-reviewed journals; and it has been found to be "generally accepted" by scientists in the field.

While that all sounds quite reasonable, the definition of what's "valid" acknowledges none of the many dimensions of judgment implicit in any one of those factors. According to David Caudill, a professor at Washington & Lee University who has studied and written extensively on *Daubert*, any acknowledgments of human judgment, differences in worldview or disciplinary approach, economic or other types of bias or conflicts of interest are "downplayed to the point of irrelevance" in *Daubert.* Because the justices wanted to provide unambiguous criteria for testing scientific ideas in courts of law, their falsely idealized notion of science won a place in the law over a broader, more accurate definition — one that acknowledges value judgments and disciplinary bias and contradictory results as perfectly typical of the practice.

As a result, those who want to challenge the purveyors of "valid science" to a courtroom debate on the risks of genetic engineering are likely to have a mighty long row to hoe, if they can get to the courtroom at all.

≡

Decisions about risk and hazards have to be made in spite of how little we know about the science in question and its consequences. But once you start to peel back the layers of how things really work in this world of technological risk assessment, see how official decisions are made and the degree to which a certain kind of expert controls the conversation and the outcome of that conversation, it makes a much harder and more fundamental question inevitable: *On what basis, exactly, should we believe our official arbiters of risk?*

For example, is our health in danger, or the health of any other living organisms, as a result of the bits of antibiotic resistance we've been digesting with our *Bt* cornflakes and our Roundup Ready soy milk lattes over the past decade? Some experts say yes, backing up their claims with studies that would almost certainly pass the *Daubert* test for "valid science." In the U.S., however, the official answer is "no," and the experts who have said so called on a different battery of scientific evidence to make theirs the winning case.

The same situation exists for virtually every topic of debate about the safety of transgenics. But as with nearly every other issue about transgenics raised here, there has been *only* debate: polarized, fundamentalist and evangelical in nature and tone. Neither "side" has yet engaged in a serious conversation that respects, acknowledges and accounts for the various legitimate perspectives on this issue, nor for the kinds of values, biases and conflicts that color the perceptions and judgments of any expert or stakeholder, no matter what position they take.

So, given how the deck is stacked, where would you lay your money down? Not that it really matters. Unless the rules of the game change, it's pretty much a sucker's bet.

CHAPTER 11.

PUTTING PIGS TO THE TEST

The point is not that we should reject science and all its works because experts disagree, or because their decisions are influenced by judgment. Uncertainty is still the most pervasive problem in risk assessment, decisions still have to be made, and judgment is pretty much the only tool we've got. But the judgment of any one person or discipline simply cannot "prove" the wisdom or relevance of a decision about any given risk. Without a more inclusive process, something of importance always gets left out. Over and over again, the criticism of risk decisions that are made or solicited by governments is that the people doing the analysis didn't really address or listen to the concerns of their constituents — that is, the people who were most interested in or affected by the decision.

Risk scholars figured this out more than a decade ago. The problem we've got today isn't a failure of analysis per se. It's a failure to allow more people to participate in the risk assessment process, people who inevitably have important knowledge to contribute. The problem is that decision makers and scientists and technologists have failed to enlarge the conversation about risk, either because they don't understand risk or because they're afraid of losing control of the outcome, or both. Given what's at stake, this is no longer a defensible stand.

"Many different kinds of policy questions are raised by problems like genetic engineering, but the way that most risk assessment is done today doesn't do them justice," said Paul Stern, who was study director for the seminal National Academies study *Understanding Risk: Informing Decisions in a Democratic Society.*

For Stern and for many of the people who served on the committee he convened, the publication of *Understanding Risk* was a transformative experience. It's not hard to understand why. It threw down the gauntlet and stated flat out that the results of math-based analytical approaches to risk and innovation are no longer acceptable on their own. Risk assessment is too subjective to be calculated, it said. It is a political, ethical and values-laden activity, period, and it needs to be conducted with full participation of everyone who stands to be affected by the decision.[1]

Because when it's not, it's impossible to know whether to trust the results. For example, when the EPA does a quantitative analysis to make a risk assessment, "they get some toxicologists together, some hydrologists, then they create some models together and think what they've been doing is objective analysis," said Stern. "But actually they've made judgments at every step — about the model itself, about variables, and about which data to include and exclude from the analysis. They're actually doing deliberative, collective judgment-making, and in my judgment, they're not the right collective to be making those judgments. At every step of the way, reasonable people could disagree. And there's no analytic technique you can count on to resolve those disagreements."

I met Stern because I wanted to find out more about *Understanding Risk.* "That report was the most important thing I've been involved with in 23 years at the Academy," he said. The report was published in 1996, and its original goal was to find a way to translate the output of a technical, quantitative analysis into a document that an audience of non-technical decision makers could understand.

"Our view was that the audience for a risk analysis is more complicated than that," said Stern. "We didn't want to buy into the assumption that output of a scientific analysis was equivalent to what decision makers needed to know. So we took a step back. We asked, how do you find out what the relevant information is — for all the interested and affected parties, not just the agencies? Quantification is one method, but what are the others? And off we went."

The committee that Stern convened to produce *Understanding Risk* included some of the most highly regarded science and technology scholars in the world, as well as epidemiologists, public policy and public health specialists, philanthropists, economists, biologists and lawyers. Those who participated and whose research helped form its foundation include several names you'll recognize from these pages: Nell Ahl, Harvey Fineberg, Baruch Fischhoff, Warner North and Paul Slovic.

Over the course of the study, the committee heard presentations given by specialists from the USDA, EPA, FDA, industry organizations, local governments, and the U.S. Nuclear Regulatory Commission. They discussed issues of nuclear waste and nuclear power, toxic chemicals and ecological risk. They reviewed the work of the leading scholars in the field. And at the end of two years of deliberation, the *Understanding Risk* committee came to the conclusion that even if all the uncertainties about any given technological risk were satisfied — in other words, even in the practically impossible presence of complete data — the process and the results of analysis, just like the process and results of scientific discovery, would still be subjective.

And so it became of central importance that the concerns of interested and affected parties be addressed in a way that was understandable and accessible to them, no matter what their level of technical or scientific expertise.[2]

The committee dubbed this participatory approach the "analytic deliberative process." Far more provocative than its thudding name implies, the process elevates the "value" side of the risk equation — call it logic, judgment or deliberation — to a role where it's just as important to a defensible analysis as any technical data or calculation. And while some technical experts continue to bemoan what they believe is the loss of scientific rigor that the approach seems to represent, the committee's work showed that the combination of analysis and deliberation actually *improves* the role of science in risk considerations, making it more directly relevant to the problem.

Instead of ignoring important areas of scientific uncertainty specifically because they are uncertain, which is what many regulatory agencies and scientists do today, they should be paying even *more* attention to important areas of study where there isn't enough data. This kind of "scientific management of science," as Fischhoff has called it, can deliver a real research agenda for *risk* rather than for scientific discovery, flagging and ranking critical uncertainties based on both available knowledge and what decision makers need to know in order to make an informed decision.[3]

=

Collaborative risk assessment — combining available data and including all the people who are interested and affected by the risk — is a road map to enlarging the conversation about risk, yielding exactly the kinds of results we need to move forward with a more sensible approach to transgenics. The *Understanding Risk* committee was the only group to say so as forcefully as it did; as a result, it provided the momentum for many other influential groups to step forward and sanction the approach. The 1997 report of the Presidential Commission on Risk endorsed the same general method, as did the Canadian Standards Association, the Organization for Economic Cooperation and Development[4], and, in the U.K., Her Majesty's Treasury.[5]

The combination of analysis and deliberation that *Understanding Risk* championed is already known to have rocked loose very old, very stuck conflicts. One example is the resolution it brokered between the citizens of Valdez, Alaska, and the marine oil trade after the disastrous 1989 tanker crash and oil spill off the Alaskan coast. The two sides had been engaged in a bitter, years-long dispute about what kinds of tug vessels should be deployed in the Prince William Sound to help prevent oil spills. Instead of funding the usual competing risk assessments in an effort to influence the decision, the Regional Citizen's Advisory Council, the oil industry and the government agencies involved in the decision decided to jointly sponsor, fund and support a single assessment.

The research team was composed of both industry and advisory council experts, guided by a steering committee with representatives from all three groups. Over the course of the proceedings, everyone involved learned about the technical intricacies of maritime risk assessment. One member of the research team said the process "increased our understanding of the problem domain ... The assumptions were brought out in painful detail and explained."[6] The team decided that the existing records didn't provide enough data for a proper risk assessment, so the steering committee helped them find the data they needed. As a result, one of the new tug vessels was deployed in 1997.

By engaging these formerly warring stakeholders in the process, the problem was better defined, the quality of the data was improved, and the assumptions and the uncertainties were openly acknowledged and given the thought and consideration that they deserved. Even more impressive, the final risk assessment was accepted as authoritative by everyone.

In another powerful example, the acknowledgment of uncertainty and a spirit of cooperation helped hammer out a thorny risk question for an EPA drinking-water regulation in a way that satisfied regulators and technical experts from various perspectives, as well as the people who were to be affected by the decision. The technical problem was that water chlorination, which kills many pathogens, also can react with other organic compounds in the water and yield what are called "disinfectant byproducts," some of which are carcinogens. The EPA turned to negotiation because it already knew that the ordinary rule-making process would be contentious and probably futile as a result.

There were two noteworthy components to the deliberation process. The first was that the EPA hired an outside firm to interview potential committee members and determine who should be at the table. It wanted to know, and to disclose to the negotiators, who among them had an interest in or was affected by the proposed rule, and how their biases might affect the ability of the rest of the group to deliberate. After much debate, the chlorine industry was not granted a seat on this committee, since it was "strongly committed" to its position about the safety of water chlorination. Instead, it served in a technology advisory group and stated the group's position at the first negotiating session.

The second component was that by the end of the process the committee had decided it still didn't have enough data or evidence to make some of the most important decisions that were required for it to come up with a hard-and-fast regulation. So it proposed an "information collection rule," which required those in charge of large public water supplies to monitor and test source water for disinfectant byproducts and also required certain detailed actions based on the results of these tests. This breakthrough compromise allows the rule to be modified over time, based on the results of ongoing data collection and evaluation.

In other words, to deal sensibly with the scientific uncertainty of the data, the committee built into its rule a *legislated* means by which to reduce that scientific uncertainty over time. And not only that, it also designed a rule requiring that the uncertainty be addressed *as part of the process of protecting public health*, instead of pretending that the uncertainties didn't exist.[7]

As these examples show, using a collaborative process to make risk decisions isn't about providing a forum for the public to vent its opinions on a controversial issue. The EPA's rule making wasn't even public, yet arguably it had greater impact on more people than the very public process in Valdez. The goal of analysis and deliberation isn't to have our governments pat us on the head and say, "We're so glad you got that off your chest," and then go ahead and do whatever they want. That seems to be the unfortunate role (and impact) of most public hearings and "citizen input" venues, as was clearly the case with the "GM Nation?" debate in the U.K. that I mentioned early on.

Instead, the most important benefit — and the whole point, really — of going through the elaborate machinations of such a process, is that by working together, the people who come to the table *improve the outcome*. They question scientific judgments and assumptions from a fresh perspective. They challenge each other's biases. They see critical uncertainties that are invisible to or ignored by experts. They help scientists come up with new research agendas to answer specific risk questions. Collaborating on risk decisions *uses our values and our judgments and the variety of our experiences as a positive force* — not only for society but, notably, also for science.

Expanding the risk conversation to include the people who will ask these kinds of questions — *before* laws are passed and *before* products are put on the market — could have a huge, positive impact on national economies around the world. Product developers and regulators could move away from expensive, risky innovations and practices to alternatives that are safer, more acceptable and thus more profitable. How much did the Flavr Savr failure cost Calgene? How much more will Roundup Ready soy end up costing the people and the government of Argentina? What will *Bt* cotton cost India? What might be different today if anyone had bothered to ask some hard questions about risk and benefit, questions that might have steered companies to less risky, more profitable strategies?

Transparency is the other critical benefit to the process, and one that seems particularly important in light of the kinds of conflicts that are being exposed between government regulators, industry and academic researchers.

"If you have a regulatory system that enshrines a collaborative, analytic and deliberative process like the one we proposed, you've created an institutional structure that works directly against outside influence," said Stern. "It's designed to figure out the information that the group needs and provide the checks and balances to prove the information is trustworthy, since all the stakeholders have the ongoing opportunity to question each other and resolve disputes."

How you actually combine analysis and deliberation in practice depends on the individual problem, said Stern. One approach uses a decision aid — called a "utility analysis" — that's well known in many disciplines, from economics to engineering to psychology. A utility analysis is a way to assign importance to various outcomes without assigning them a monetary value. This avoids some of the more obvious problems with cost-benefit analysis by assessing many different criteria and ranking their varying levels of value (i.e., utility) to the participants.

"Imagine that you're going to make a decision about whether an experiment should go forward, or a technology should be allowed to go to market," Stern explained. "You start by asking a bunch of people you've selected, reasonable stakeholders with clear perspectives on the subject, 'What are all the possible outcomes of this experiment that would matter to you?'" Let's say that for some people, what matters is making a profit. For others, it's losing or gaining a habitat for a species. Yet others want to stay healthy. You ask them to place a numerical value on the outcomes that matter to them.

"Then you go to the analysts, give them the list and ask them, 'What can you tell us about the effects of the decision on each of these outcomes?'" said Stern. "You'll get lots of different answers based on everyone's value structures, the analysts' values included. But in any case, at that point each stakeholder could take the answer they get from the analysts, multiply it by the value they placed on their most important outcome, and get an answer of either 'Great, move forward,' or 'This is terrible, stop.'"

In short, you find out what people value, you get the best possible information on how the experiment will affect what they value, and you discuss it until you can see a path to a decision. Thankfully, not all risk problems require such an involved approach. But we should be equally thankful that it exists for the problems that do.

"You really only need to use the analytic deliberative process under certain conditions," said Stern.

The first instance is when lots of different dimensions can be affected by the outcome — like a market or an economy, health, the environment, an endangered species, and so on. Second is where there's scientific uncertainty — where there's not enough science to know how things will turn out in the long term — and you need to make a decision before you know. Third is a problem where there are value conflicts embedded in the uncertainty — that is, where there are disagreements about benefits or about which outcomes are important, and people's minds may change if they're given new information. (Think "GM Nation?") Fourth is where there are no authorities you can trust to know all the answers. And fifth, he said, analysis and deliberation are important where there's urgency, where you can't wait for the situation to "get certain" before you make the decision.

The risks of genetic engineering meet all five of those criteria, and according to Stern, "that's why it's a really good target for the approach."

≡

I was curious to see these benefits for myself. So Baruch Fischhoff and I took a proposal to the National Science Foundation, where we were awarded a small grant for exploratory research. We wanted to see if we could develop a method, founded on the tenets of analysis and deliberation, that would work with a risk problem of extraordinary difficulty and consequence. It's a risk that is holding back full-throttle development of a human intervention that could save many lives: the use of transgenic pig organs for human transplants, a technology known as xenotransplantation.

There were a couple of reasons this problem seemed so intriguing to us. First, the possibility that a dormant retrovirus from the pig's cells would somehow reactivate inside the human body poses a risk that's literally incalculable. Living pig cells have never before been able to actively exchange proteins and genetic material with cells inside a human body. The result of a viral recombination or crossover from pig to human would likely be a unique, unpredictable and potentially deadly disease never before encountered by the human immune system. (The pig and cow heart valves that are used in hundreds of thousands of human patients are no longer living tissue.)

We also liked the idea because it was controversial in dimensions beyond the technical risks of the xenotransplants themselves: several of the biologists in the group had dismissed the technology entirely. "Stem cells will be able to grow new organs before it gets to that," they said. And while they may yet be proven correct, not long after our initial conversation, a story in the *Wall Street Journal* told of an announcement by the South Korean government that it would spend 60.5 billion Korean won (about $62 million) over the next 10 years to perfect the mass production of pig organs for human transplants, as part of a government-backed push to excel in biotechnology.[8] So we had found a problem where scientific controversy was coupled with deep scientific uncertainty.

What useful new perspectives could a diverse group of experts bring to that conversation?

To find out, Fischhoff and I assembled a team who spent nearly a year building the technical foundation for a deliberative process: writing scenarios and designing probability models for the various ways that a porcine retrovirus could jump to or recombine with a human's. When they'd finished, we invited half a dozen experts from various fields to critique and expand on our results.

And within the first half hour of the meeting, our flat model about the risks of a controversial medical procedure took on some surprising dimensions that none of us had foreseen — far beyond the scientific facts and uncertainties that it represented.

"What do you do with the carcasses?" was the kickoff question of the day, from Kim Waddell, who had been study director for the *Animal Biotechnology* report. "For xenotransplants to be affordable from a commercial standpoint, they're going to have to be done at a scale that actually generates serious income," he said. That means slaughtering thousands of animals, and not just those whose organs will be used. There are bound to be plenty more that are rejected for some reason — their gender, perhaps, or they were born with the wrong suite of genes.

"Doesn't matter whether you harvest their organs or not, all transgenic pigs are considered contaminated," he said. "So you can't use them anywhere in the food chain. Animals can't eat them, and neither can we. You can't render them. You can't bury 30 or 50 or 200 pigs in someone's backyard. And you can't burn them. That's an ecological risk as well."

Also, he said, someone has to think about the fact that the pigs will almost certainly enter the food chain anyway. In fact, that has already happened: transgenic laboratory pigs at the University of Florida were marked for destruction, and "someone backed up a truck, loaded up the pigs, ground them into sausage and sold them to his friends," Waddell said. "So we may get the 'Jimmy Dean' solution whether we like it or not."[9]

Alwynelle "Nell" Ahl was next up with similar concerns. Ahl is a zoologist, a biochemist, a veterinarian, and a longtime administrator for the USDA, whose last job there before her official retirement in 2000 was as the founding director of the USDA Office of Risk Assessment and Cost-Benefit Analysis. She was one of the advisors to the *Understanding Risk* study in 1996. Her concerns were about a more specific byproduct. "Remember that when you raise pigs, they produce manure and I mean in huge volumes," she said. "And you've got to do something with it, because it can be a source of genetic material that's shed from the intestinal lining. If the pig's transgenes are considered a contaminant, then when the manure leaves the pig, another big problem is rodent control, and insect control in particular."

Biting and sucking insects pick up pig genes, whether from manure or flesh, and transfer them to other organisms. So not only are we going to have to worry about getting rid of the pig carcasses, but we're also going to have to be concerned about what we do with the manure that is produced. "It's not just the disposing of manure, but also sterilizing it," said Ahl. If you need to go to scale to make xeno economically viable, the concern for manure would be a very expensive one."

Waddell also pointed out a risk that wouldn't necessarily be obvious to someone assessing human health concerns. "Pigs are social, gregarious and highly intelligent animals," he said. "Early indications are these animals are going to have to be isolated to keep them from contact with potential infectious agents, and so you're talking about the potential for some pretty unhappy pigs. Unless you hire people to dress up like pigs to entertain them while they're in isolation, they aren't going to be getting enough stimulus. Unhappy animals are stressed animals, and stressed animals get sick. And it raises the issue, how healthy will these animals be for the purposes that we have intended?"

Other questions were raised about what kind of pig had been selected to undergo the "knockout" procedure. Is the particular species that's being used to propagate the "knockout" modification more or less susceptible to certain other kinds of diseases? And let's say that we have managed to sanitize a pig in some way. Its organs maybe aren't carrying so many viruses — maybe they're even approaching sterile. What happens when you put a relatively clean pig organ into a dirty human body with all the viruses we carry around? Could the human diseases kill the pig organ?

Margaret Cary, the medical doctor of the group, as well as a business administrator for the Veterans Administration in Washington D.C. and chair of the University of California, Berkeley, School of Public Health's Policy Advisory Council in 2005 and 2006, said that "dirty" humans do this today, with human organ transplants. "We're already retransplanting hepatitis B patients because of this, so sure, it could happen with xeno," she said. "And in that case, you'd just have to get a fresh pig and do the whole thing over again."

Xenotransplantation isn't likely to be the only organ replacement-type therapy that will end up being used by humans, and one panelist saw that fact as presenting another potential risk.

"I actually do think that the tissue that's engineered in stem cell approaches will eclipse these xenogenetic methods. But what will probably happen is that there will be a combination of different kinds of cellular therapies," said Linda Hogle, a medical anthropologist at the University of Wisconsin whose research focuses on tissue engineering, transplant medicine and emerging medical technologies.

"I think what will happen is that there will be a two-tiered system. People who can afford it will get stem cells or hybrid devices," Hogle said. "People who can't afford it, or poor countries, will get the xenogenetic materials. Then you might see a stigma about contamination and people who are potentially 'dangerous' and 'risky.' And then we're talking about issues of class, not availability."

Hogle then raised an even more frightening possibility from the public health perspective: how do you control what people do after they have the operation, and who may be walking around with viruses in their genomes that are waiting to be triggered by some random or stressful event?

"If you start with a situation where the animal materials are all that's available, it's probably going to be the sicker people — the ones who don't care as much about the risk — who will get them, right? They're not only going to be in worse condition at the time of the procedure," she added, "they'll probably be less likely to have ongoing checkups, or submit to ongoing surveillance to see if they've contracted the virus over time, or keep their promises to use contraceptives to prevent any genetic changes from being passed on. So they may do worse after the operation, *and* be more susceptible to both getting and spreading an infection than if you were doing this in a known healthier population of people with good support care."

Sally Kane, an economist with 20 years of experience in research and research management with various U.S. agencies and legislators, including the USDA, the Department of Commerce and the National Science Foundation, flagged another important potential public health risk: patients traveling abroad to get cheap transplants, who then return to their own countries, communities and families, harboring the potential for a retroviral infection.

The discussion continued along these most enlightening lines: quality-control issues about the organs themselves, the inevitability of "transplant mills," what to do about people passing along the potential retroviruses to their children conceived post-transplant, and the new legal liabilities that might result when organs are no longer part of the "gift" economy as they are today but instead can be purchased as "products" or medical devices. When we ended the day and went back to our disease model, it seemed possible, based on the contributions of our panel, that the scientific risk factors for contagion by porcine retroviruses might be increased by the social and economic factors that could influence how xenotransplants were practiced in the real world.

These were the issues that came up over the course of less than one day. At the risk of stating the obvious, this is most certainly not the kind of result you would *ever* get out of a traditional risk calculation about the probability of a species-hopping virus. But it sure is what you'd hope some decision maker would think about before deciding to approve transgenic pigs as commercial products for transplants.

Despite the fact that scientists continue to be wary of this kind of collaborative, value-driven process, a growing number tend to acknowledge — even if sometimes begrudgingly — that both judgment and analysis are required to assess risk in a useful way. It's the most important thing that has happened in the study of risk over the course of the past 30 years: many more scientists are now convinced that value judgments, contingencies and uncertainties, and their own inherent biases and omissions, all affect the outcomes of any kind of scientific assessment, not just the assessment of risk.[10] Science has a vital contribution to make to this process, of course, but the "evolving insight" for me, to paraphrase Fischhoff, is that for these kinds of complex problems, science is just one input — necessary but insufficient.

That has always been the paradox of this diehard connection that has been forged between risk and science. Governments and regulators have given scientists de facto authority over risk decisions, with much of the planet's health and well-being at stake if they're wrong — even though testing each other's results and proving each other wrong is what scientists do. In the daily practice of science, in fact, most experiments fail. Even when one succeeds, whatever scientific "truth" it has produced is a moving target; researchers around the world are working hammer and tongs every minute of every day to disprove each other's work and come up with something new of their own that will, in turn, earn the privilege of being subjected to the anvil of scrutiny.

But even so, experimental results don't exist in a vacuum. Once you've factored in the politics of research and regulation and funding and all the other things that have been proven to influence the practice of science and the "truth" it produces, our reliance on scientific evidence as the official arbiter of risk strikes me as not just unfair to people like us who must live or die by its decisions. It's also unfair to the scientists, asking them to bear the brunt of responsibility for anticipating the consequences of forces that they understand only within the very narrow range of their expertise and unavoidable bias. Hobbled as they are by the human traits they bring to the work they do — hobbled as unavoidably as we all are by our values and prejudices — we've ceded to them an unjustifiable and sometimes dangerous authority over many of the risks of modern life.

Enlarging the conversation as we began to do in Berkeley is not itself without risk. If it's not done with complete transparency and a real desire to both understand and make a decision about the risk at hand, these kinds of processes might cause innovation to spin into a twilight zone of endless debate, with some of the potential benefits of biotechnology never making it past the conceptual phase. But this is a pretty unlikely outcome. Of course there are some people who believe that any human intervention into biological functions is wrong, just like there are some people who believe that any biotechnology scientists can cook up and that doesn't drop people dead in their tracks will work out fine in the long run. But most people are somewhere in the middle. They want to live longer, healthier lives and are happy to accept the risks of new technologies — as long as they trust that someone is truly paying attention to what might go awry before all the kinks are ironed out.

Even better, the deliberations can be so enlightening from the scientific perspective that it's hard to make a case for *not* doing them. That's what was so striking to us, after having convened the xeno meeting. Our panelists claimed to be greatly informed by the process itself and were unanimous in their clamor for more of that kind of conversation. Even though most of them were trained in one of the natural sciences, they quickly saw the gaps in scientific knowledge, and saw as well that the science alone wasn't very helpful to them from the perspective of making the decision at hand, about whether xenotransplantation's theoretical benefits might pan out in the real world.

It was the whole picture, they said, including as much of the landscape as possible, that was useful to them. Knowing all of the potential variables, weighing the real risks against the real benefits, was the only way they could feel like they were making a responsible decision.

"Let me just add a little bit of perspective, about why this approach is something that interests me," said Nell Ahl. "Problems like xeno and other biomedical and gene technologies are fusing into things that are going to have pieces in every part of government, and right now there's no way to get people together to work on them." For example, at the USDA, the people on her team would find themselves working on a problem where a piece of the risk actually came under the FDA's jurisdiction. "But the FDA wouldn't talk to us about it. Another piece belonged to the EPA and they said, 'Ag Department? Who are you? We don't like you guys.'"

"And so models like this one, that bring disparate groups together on common problems — I see this process as something that would give each of the agencies a seat at the table and a reason to get together," she added. "Maybe that's being idealistic, but that's why I'm hoping that something like this can come to fruition. Each agency makes a decision on their little slice, and there's no one looking at the big picture."

CHAPTER 12.

WHAT THEN SHALL WE DO?

The big picture is the only way the world makes sense to me. In fact, when I started doing the research for this book, I thought the best approach would be to create a big picture of my own: to make a map of what was known and what was unknown about the science behind genetic engineering. I wanted to chart the peaks of knowledge about molecular biology and genomics and the valleys between them, to make something tangible that I could point at and say, "Ah, here is gene silencing," for example, "but over there is what scientists would need to know before they could safely commercialize it."

If such a map existed, I thought two things would become obvious: one, that trying to use scientific evidence alone to effectively forecast risk was the ultimate fool's errand, and two, that the map itself could reveal a path, for scientific discovery and for the social acceptance of risky biological innovations, that would be of tremendous benefit to researchers, regulators and the public who must live with the consequences of their decisions.

But the knowledge landscape was evolving too fast in too many dimensions, so at some point I abandoned the map idea. It wasn't until we'd finished the xeno meeting in Berkeley that I realized we'd done it. We'd created a map of knowledge — of certainties and uncertainties, both social and scientific — in a form that anyone could understand and contribute to, from subject experts to people who had no special training at all. We certainly hadn't mapped everything about xenotransplantation, let alone genomics, but we'd gotten an idea of the shape and size of the problem. Better yet, we'd started building a toolkit for anyone who wanted to assemble the requisite experts and stakeholders and build their own uncertainty maps, one risk problem at a time.*[1]

Still, Ahl's final comment about warring regulators brought my congratulatory musings about our meeting back down to earth with a thump. The truth was unavoidable. No matter how thoroughly we mapped the uncertainties of any scientific innovation, no matter how many experts or stakeholders we invited to take part in the risk deliberations or how persuasive our results might be, none of it would matter unless the methods were used by the agencies that regulate risky technologies.

Just as importantly, they had to be used early enough in the process to avoid the kind of no-win situation that she described, and which exemplifies today's "debates" about transgenics: where too much turf, money or ego is at stake, and people simply won't budge from their positions once they've become invested in them.

So the question became not just how to better assess risk, but how and where in the halls of power to install this kind of transparent, open deliberative process so that its results could actually affect the outcome.

One possible course of action that occurred to me was to reanimate a 21st-century "biotech" version of the much-mourned OTA. The OTA, formally known as the U.S. Congressional Office of Technology Assessment, was voted out of existence as "too costly" by a conservative Congress in 1995. Yet in hindsight,

---

* In fact, later that year we expanded the capabilities of our 'toolkit' to create a knowledge map of the avian flu pandemic, including social, economic and scientific uncertainties. The results were published in the *Journal of Risk and Uncertainty*.

many lawmakers and policy experts of all persuasions have begun to acknowledge that eliminating the OTA was a mistake. For more than two decades — starting in 1972, just before recombinant DNA was invented, and ending just after the commercialization of the Internet — the OTA had provided them with unbiased advice, assessments and policy analyses on an extraordinary array of issues and advancements. So far, no other office or agency in the federal government has been able to fill its shoes.

Forged in the crucible of partisan politics, the OTA was widely respected for its political neutrality; it was equally accountable to Democrats and Republicans, as well as to various congressional committees whose jurisdictions overlapped or competed.[2] Its constituents included not just politicians, but also scientists, stakeholders and other subject experts who participated in its assessment processes. Much of the OTA's work was in response to specific requests from lawmakers about technologies that were already available or just around the corner. But just as importantly, OTA reports were often ahead of their time, assessing the impacts of new technologies or nascent trends at their earliest stages, long before they reached the market.[3]

It also could be fearless about looking at the big picture, regardless of whose ox might be gored in the process.

One example from 1992 involved the OTA being asked to evaluate a technology called "difficult to reuse" (DTR) needles for drug injections. The group requesting the assessment, a subcommittee on regulation and business opportunities, wanted to know whether legislating the use of DTR needles would reduce HIV infections in the high-risk population of injecting drug abusers.[4] OTA concluded that most of the proposals for DTR equipment not only would be unlikely to reduce the spread of HIV, but would probably have other negative indirect and social consequences as well; i.e., creating a black market for reusable needles, imposing a significant financial burden on injection-dependent diabetics and sharply increasing the volume of contaminated waste, thus increasing the cost for its disposal.

Clearly the OTA's analysis had ranged far beyond a simple evaluation of the needle "technology" itself. It had incorporated both how the needles might be used by drug users, as well as its broader social impact — including, but not limited to, the purported benefit of preventing the spread of HIV. As a result, despite the requesting committee's obvious predilection toward creating a market for the new needles, the report was persuasive enough to cause the legislation to be withdrawn from consideration.

We could be well served by reviving an OTA-like organization for the purposes of biotech policy assessment. We might call it something like "BIOTA" — which, in addition to being an obvious mash-up of "biology" and OTA, is also a term for all the living organisms of an ecosystem. BIOTA could cast a fresh eye on a regulatory regime for biotechnology that has been held captive for too long by special interests and blinkered scientific perspectives on risk.

With an official recommendation now on the books that someone ought to be monitoring and tracking transgenic food crops, for example, BIOTA might investigate how best to turn that recommendation into law.* This would probably require a technical assessment of the least expensive and most effective methods to get a monitoring and tracking system up and running properly. BIOTA might also recommend that some agency, possibly the National Science Foundation, solicit proposals for a more sensitive device to detect and monitor transgenes in the field, an

---

* One idea that Roger Brent and I came up with several years ago was a non-governmental Bio Peace Corps, to give young molecular biologists, in particular, the opportunity to leave their labs and head out into the field. Using the most advanced detection technologies, they could sample animal and plant populations for transgene flow and monitor air, food and water supplies for transgenes, transgenic proteins and other biotech products or artifacts. Some of them might be able to take on a role in national defense, as participants in a "Biological NORAD" that monitors the environment for incoming biological agents in the same way that the North American Aerospace Defense Command looks for incoming airspace threats.

activity that is expensive, impractical and not very effective with today's technologies. It might also prompt a long overdue evaluation of liability issues for transgenic food, should safety issues present themselves.

BIOTA could also re-evaluate the present patchwork of government regulations for transgenics, the classic example of which Nell Ahl provided at the end of the xeno meeting. It could recommend a new approach that's less likely to perpetuate agency turf wars, while doing a better job at properly assessing risk.

Another issue of critical importance that BIOTA might take on is the dangerous disparity between the generous funding that's available for discoveries in the biological sciences, relative to the paltry resources available for research into issues of risk, biosecurity and biosafety. Recall that less than 2 percent of the USDA's 2003 budget was routed to its Biotechnology Risk Assessment Grants Program; this outsized emphasis on discovery is not unique to the USDA. Other agencies, including the big funders like the NSF and the National Institutes of Health, follow largely the same formula. As a direct result, scientists whose primary interest is risk and safety assessment are considered second-rate by many in the research community, in addition to being cursed in terms of financial support.

An unbiased appraisal of this imbalance, and how it might well make us more vulnerable to various kinds of biohazards, might bring more well-deserved and overdue funding, researchers, attention and scientific discoveries to the critical field of biosafety and biotech risk.

Perhaps most importantly for the long term, an organization like BIOTA could become the convener for the kinds of deliberative risk-and-benefit biotechnology assessments that have been the focus of this book. I can envision lawmakers funding specific BIOTA working groups, in the same way that today they can ask the National Academies to study a certain issue or technology. If national lawmakers aren't paying enough attention, state and local governments might see the benefit of starting their own BIOTAs. And as the media and the public learn about them, BIOTA reports could become increasingly important considerations in the policy process, as long as they're maintaining their reputation for neutrality, inclusiveness and transparency.[5]

In short, BIOTA could be the canvas on which we paint the big picture: a place where regulators come together with scientists and stakeholders to deliberate and make decisions about the risks and benefits of these critical technologies, and the social and scientific issues they raise.

≡

No matter what form this trusted overseer takes, we need it now. "Done is done" may be the verdict on policies for existing transgenics, but science and commercial investment are hurtling forward in other areas that will require far more thoughtful attention to risk than anything we've seen to date.

In fact, two powerful new, unregulated biotechnologies are already gathering momentum. Both are heavily funded; one has already entered the global marketplace. Each is at least as radical as anything in the world of transgenics. So far, their supporters are lining up to be even less enlightened about risk than those who delivered today's avalanche of transgenics to the market. And we are wholly unprepared to deal with them.

The first is nanotechnology, a way to precisely control individual atoms and molecules on an unimaginably small scale, in the range of a billionth of a meter. In a move that recalls the biotech frenzy of the 1980s, billions of private research dollars are being invested worldwide to create a new class of nanoscale materials, structures and devices. The U.S. government also has launched a heavily funded National Nanotechnology Initiative to provide an early boost to what it has projected will be a $1 trillion global marketplace by 2015 for goods and services using nanotechnologies. Medical researchers in the field believe that nanotech will someday allow them to regenerate or restore any organ or tissue in the body including eyes, ears and nerves, as well as revolutionize drug delivery and disease diagnosis.

But serious health and environmental hazards are already being regularly documented, both in the lab and in the market.

For example, researchers have found that one type of nanomaterial called a buckyball, being tested to deliver drugs as well as to provide the scaffolding for rebuilding skin or bone inside the body, has damaged brain cells in experiments with fish.[6] Cosmetic manufacturers have long since re-engineered titania, the white pigment used in paint, into nanoparticles that are ingredients in many, if not most, of the transparent sunscreens on the market today. Yet titania nanoparticles are capable of crossing the protective blood-brain barrier, resulting in the kind of stress to brain cells that is thought to be the cause of diseases such as Parkinson's and Alzheimer's.[7] Other nanoparticles can migrate from the nose to the brain and from the lungs to other parts of the body; in 2006, for example, a German company was forced to recall its Magic Nano sealant after several people using it reported breathing problems, some requiring hospitalization.[8]

Researchers have known for years that nanoscale molecules behave differently from ordinary-sized molecules. Yet hundreds of nanotechnology-based products are on the market, governed only by existing regulations for toxic chemicals, which clearly have fallen short of protecting consumers from harm.

Some longtime nano-watchers are trying to raise signal flares about this dangerous lack of oversight. The Project on Emerging Nanotechnologies (PEN) published a report in 2006, outlining a comprehensive research agenda for addressing nanotech's risks. "Certain applications of nanotechnology will present risks unlike any we have encountered before," it asserted, yet only 1 percent of the many billions of federal dollars invested in developing nanotech have been directed toward risk. The report's authors tried to inventory the risk-relevant work that was being funded by the government, with limited success. "The inventory is not comprehensive," said the report, "as a number of [regulatory] agencies were not forthcoming in providing information."[9]

How scary is that? If regulators won't even *talk* about the risk research they're doing (or not doing), I despair for PEN's critical efforts to go much beyond the report stage — unless, of course, there's a nanotech-driven catastrophe that forces the agencies to take action, or until someone can convince them of the potential benefits of a deliberative process and bring them voluntarily to the table.

Until then, according to PEN's director, David Rejeski, who also directs the Foresight and Governance Project at the Woodrow Wilson International Center for Scholars, nanotech is "the perfect storm: lack of trust between all the parties, lack of consumer awareness, not enough money to assess the risks, and an industry that's nervous as hell, but not nervous enough to stop development."[10]

Following perilously close on the heels of nanotech's example is the emerging discipline-cum-industry known as synthetic biology. Called "genetic engineering on steroids" by its critics and "constructive" biology by its proponents, synthetic biology's ambition ranges far beyond merely recombining the DNA of various existing organisms. Its disciples are developing a technology they call "DNA synthesis" — that is, the ability to construct gene- and genome-length DNA fragments from scratch. They say that DNA synthesis will allow them to assemble and animate wholly new organisms, as well as to redesign existing organisms "for useful purposes."[11] The pollution-munching microbes that Craig Venter wants to invent, for example, will be the products of his new company, Synthetic Genomics, Inc., founded in 2005.[12]

Given what you've read so far of how little is known about even the most basic components of "natural" biological systems, any serious development effort in the realm of synthetic biology may seem dangerously premature. Yet the cash is already pouring in. In fact, in 2005, the first synthetic biology company had already raised some $13 million in venture funding from some of the leading investors in Silicon Valley — despite the fact that the founders of the company were, at the time, virtually the only academics working in the nascent field.[13] Not surprisingly, as with nanotech, U.S. taxpayer money is starting to flow to the field as well; in August 2006, for example, the NSF funded a $16 million synthetic biology center on the University of California, Berkeley, campus.

Both nanotech and synthetic biology advocates are well aware of the public's ongoing negative reaction to genetic engineering and, despite the stonewalling that some nanotech proponents have engaged in to date, both have professed a desire not to make the same mistakes as the genetic engineers did back in the early days. But it's important to note that the mistakes they seem most eager to address have to do mostly with the public's *perception* of risk, not the reality of the risks themselves.

For example, the intellectual core of the synthetic biology world convened its Second International Meeting on Synthetic Biology in May 2006, in Berkeley. The group invited more than 300 researchers, government representatives, policy analysts, ethicists, lawyers, economists, administrators, educators, psychologists, anthropologists, philosophers, nonprofit leaders and journalists to discuss the current state of research and expected scientific developments.

After the meeting, in keeping with its professed desire to make sure the public is informed about the work that synthetic biologists are doing, the group posted a public declaration of what had transpired. Among other things, it acknowledged that the technology they were developing could be "misapplied" (i.e., used by bioterrorists), and resolved to have "ongoing and future discussions with all stakeholders" on the technologies they develop.[14] Yet the "multi-stakeholder survey" that the declaration mentioned as part of these ongoing and future discussions is being directed by a staff member at the Venter Institute, whose founder, Craig Venter, is not exactly a disinterested observer in the outcome.

What's more, not once in that public declaration did the word "risk" or the phrase "risk assessment" appear. Not once were the basic tenets of risk, like "uncertainty" or "unanticipated consequences" directly and openly addressed. If the above list of disciplines represented at the meeting was complete, there wasn't even a risk expert in attendance. Held up to the standards of real risk communication, this public declaration failed miserably.[15] Its intent was not to communicate with outsiders about risk; it was to present synthetic biology in the best possible light, while striking the proper public pose about being open to dissenting views. It was spin, plain and simple.

And for us, it's déjà vu, all over again.

≡

What can be done? Can we pause the action this time, long enough to confront the real risks of these radical new technologies — before they've slipped beyond our control?

We *can*, of course. We've found the method and we've always had the ability. But method and ability alone are meaningless unless we use them. As David Rejeski has said, we cannot "Google our way to enlightenment" on these kinds of issues.[16] The bigger question is, are we willing to seek out the next step and take it?

Before we consider our answer, it is worth noting note how differently the situation with transgenics has been handled outside of the U.S. Informed by ongoing, public conversations about many of the same scientific uncertainties and concerns raised in these pages, regions and sometimes even countries have banned field trials or are withholding approvals of transgenic products until they have more satisfactory data on risk and benefit. In August 2006, in the European Union alone, more than 3,400 local governments in more than a dozen countries had declared themselves "GM-free" zones.[17] Many countries require transgenic products to be labeled, and carefully monitor them in the field.

Compare this to America, the birthplace of biotech and nanotech and the most information-saturated culture on the planet. Our most noteworthy reaction to transgenics among us is that there really hasn't been one. With no product labels and no monitoring or tracking laws, most people in the U.S. don't even know the extent to which transgenics have inundated their lives. They have no idea that most of the packaged food they eat has transgenic ingredients.[18] Some of those who do know about transgenics, don't care; they believe the products have been properly tested or they wouldn't be on the market. Others are concerned, but know that the Byzantine regulatory environment in the U.S. means nothing will change until disaster strikes, so they don't bother to protest. In any case, for more than a decade there's been no noticeable consumer reaction to transgenics in the U.S.

And when consumers don't react, regulators don't act. It's as simple as that.

Unfortunately, this means the burden for risk reform in the U.S., at least, falls as usual on the few who know, who care and who want to do something about it. The good news is that, for once, we actually *can*. We don't have to launch any campaigns. We don't have to wait for regulators or industry to be forced by circumstance into changing their methods. Henceforth, we can choose never to accept another skewed or biased risk assessment. That's the beauty and the power of the deliberative process. Whether the deliberation is on transgenic vaccine development in the local university labs or the effects of nano-cosmetics on the water supply, anyone — nonprofits, universities, chambers of commerce, progressive industry groups, newspapers, churches — can host an open, expanded risk deliberation about the technology questions that matter to them. All they have to do is play by the rules. And the global Internet will let us watch them from the sidelines, so we can learn from the process as well as from its results. It's not a perfect solution and it's not policy, but it's open and public, and for the present it's far better than what we've got.

Frankly, I wasn't expecting to end up angling for political change when I started this research. I thought I was going to do a straightforward review of how the safety of genetic engineering was being determined today, note where the methods fell short, and be done with it. What I didn't know, and I didn't see it coming, was what would happen when I read the *Understanding Risk* study for the first time. Once I had grasped the full implications of its antidote to the abuse of scientific evidence and regulatory politicking, it dawned on me that all was not lost just yet. The examples I found of the deliberative process at work, as well as my own experience, showed me that even the beginnings of a more inclusive conversation about risk held tremendous promise — not just for the development of safer, more useful products, but eventually, possibly, for restoring the public's lost trust in science, industry and the processes by which we're governed.

How precisely these changes will come to pass, I don't know. But I do know they *can* come to pass, because the ideas upon which they are based are not new. They are the bedrock of democratic society. Making room for new voices and real dialogue at the risk table can do far more than produce better decisions about risky technologies. Best of all, it can provide us with tangible evidence of why practicing democracy was such a good idea in the first place: it actually works.

ACKNOWLEDGMENTS

Many remarkable people helped birth this book after a long gestation period, and I'm delighted to be able to acknowledge them at last. First and foremost, I am indebted to the many scientists and scholars who helped me to develop and clarify the arguments presented here.

At the top of the list is Roger Brent, whose prodigious love for science, humanity and a good intellectual tussle helped spark the original concept. His excellent and unstinting criticism of this work over the years gave me courage and terrified me in equal measure. I am every bit as indebted to Diana Rhoten, who started The Hybrid Vigor Institute with me in 2000, and now is a program officer at the Social Science Research Council. Diana both introduced me to Baruch Fischhoff and helped me begin compiling the research on this topic long before it became a book; personally and professionally, she has been of immeasurable value.

I am also deeply grateful to Baruch Fischhoff, who first led me to the National Academies' risk studies that have been so influential to my thinking, and whose generosity and willingness to partner with an amateur like me continues to enrich and gratify me. I also have Baruch to thank for introducing me to Warner North, who subsequently spent many hours explaining to this neophyte the intricacies of probabilism, decision analysis and, most memorably, Pascal's Wager. Thanks also to Warner for introducing me to Harvey Fineberg, who was kind enough to engage in an extended conversation about the details of the *Understanding Risk* study. His experience and perspectives on the inner workings of research, policy and risk provided answers to some of the questions that had puzzled me the most.

Many, many thanks to Todd R. La Porte, who managed to scare the hell out of me early on by asking the two hardest and most relevant questions of all; and to Paul Thurman, for his tremendous patience and creativity in explaining statistics to someone who managed to graduate university without taking a statistics course. He made a very tough subject come alive. I'm also indebted to Paul Stern, who very kindly helped me understand both the scientific and the political context in which the Academies' risk studies were conducted; to Robert O'Connor, the program officer for Decision, Risk and Management Sciences at the National Science Foundation, whose willingness to let us explore our ideas led to such a terrific result; and to Paul Slovic, for digging up and sharing such an illuminating piece of his archives with me. I also must thank John Adams and Charles Perrow, authors of *Risk* and *Normal Accidents*, respectively, and Jerry Ravetz, associate fellow at the James Martin Institute for Science and Civilization at the University of Oxford, for their kindness, patience and sharing of their important perspectives.

I owe a tremendous debt of gratitude to the biologists who worked closely with me over the course of this project: Nell Ahl, whose practical experience provided a sturdy bridge from the lab into the real world; Jack Heinemann, whose humor, ferocious intelligence and mountains of source material gave me the backbone to ask the harder questions; David Thaler, the most open-minded scientist and one of the most compassionate and generous people I've ever met; and Kim Waddell, who deserves much more credit than he'll ever get for the National Academies' transgenics studies that he directed. Every conversation was, and continues to be, a pleasure.

I'd also like to thank Elaine Bearer, a professor of pathology and laboratory medicine at Brown University, for her insightful comments on the draft manuscript; Michael Holland, a professor of biochemistry and molecular medicine at the University of California, Davis, for his cogent explanations of his work on transcription factors; and David Quist, for his provision of journal articles on transgene instability.

I'm very grateful to the members of the xenotransplantation project team who developed the scenarios and influence diagrams for the 2005 Berkeley meeting: Wändi Bruine de Bruin and Umit Guvenç, the risk and decision team at Carnegie Mellon, and Chris Armstrong, a molecular biologist now at Washington University. It was thrilling to see a raw and naïve idea I'd had so many years earlier start to come to fruition at their hands.

Many thanks also to the attending panelists I've not yet mentioned here, whose participation yielded such juicy fruit: first and foremost, Steve Weber, who hosted us and who has long been a supporter of Hybrid Vigor's work, Margaret Cary, Linda Hogle and Sally Kane. For very early inspiration, I must also thank Benjamin Kuipers, a computer science professor at the University of Texas in Austin, whose specialty is the effective use of incomplete knowledge, as well as Janet Maughan, deputy director of the Global Inclusion Program at the Rockefeller Foundation, who funded Hybrid Vigor's first white paper on assessing technological risk, and Kjell Andersson of Karita Research in Stockholm and Chris Elliott of Pitchill Consulting Ltd. in Surrey, U.K. Kjell's and Chris' work on the limitations of scientific expertise in risk assessment had a tremendous impact on me when I started down this path in 2001.

For sharing with me their important work and insights about science and public policy, I thank David Guston, M. Granger Morgan, David Rejeski, Michael Rodemeyer and, coming in just under the wire, Todd M. La Porte, who is happy proof that the apple does not fall far from the tree.

As for the making of the book itself, I'm glad I can finally acknowledge Lydia Wills, agent extraordinaire, whose faith in me and in the importance of this subject has sustained me over three and a half long years. I also heap infinite thanks and praise at the feet of my dear friend and excellent editor, Janice Maloney; without her expertise, insight and generosity, I would still be hunched and muttering over the keyboard.

I'm also grateful to several very talented friends: Nathan Shedroff for every one of his terrific cover designs and photos, as well as for his extraordinary patience and generosity; Lisa Raleigh, who first started dotting the i's and crossing the t's from a perch in faraway Oregon; and Erfert Fenton for her willingness to take on the unenviable job of indexing this tome.

Many thanks also to Jamie Shor and her team at Venture Communications in Washington, for knowing what to do with a book like this; and to Lulu.com for the terrific service it provides, as well as its support of this project. I'm also very grateful for the indispensable skills of transcriber Janet Vanides, as well as copy editor Sharon Gostlin and attorney Eric Rayman, both of whom signed on to help without even meeting me. Thanks also to Marjorie Baer, Victor Gavenda, David Van Ness and Stephan Somogyi for vital, just-in-time PDF, Word and font advice; and to Mike Whistler, for his phenomenal programming skills and ongoing support of this project and Hybrid Vigor.

Boundless thanks also to those who read and commented on various pieces or versions of the manuscript, especially Lynne Bolduc, Michele Di Lorenzo and Lisa Gansky; I believe they've read every word of every iteration since Day One. Also to: John Battelle, Mark Benerofe, Deborah Branscum, Joel Garreau, Gerry Howard, Roxane Johnson, C.J. Maupin, Oliver Morton, Norman Pearlstine, Marni Rosen, Sally Rosenthal, Lisa Simpson, Kerry Tremain, Elbert "Doc" Tuttle Jr. and Stacy Williams.

In addition, I'm beyond fortunate to be surrounded by a tremendously supportive group of friends, colleagues and family: My mother and father, Marge and Tony Caruso, David Baron, Sunny Bates, Anna Bisson, Andrew Blau, Girija and Larry Brilliant,

Cheryl Brooks, Kimi Burton, Anita and Dana Cadonau-Huseby, Doug, Gordon and Sophie Camplejohn, Maggi Cary, Cathy Cook, Chris Desser, Tracy Di Miroz, Mark Dowie, Robert Gatto, Daniel Goldman, Matt Heckert, Steve Hayden, Don Hazen, Katherine Fulton and Katharine Kunst, Jim Hayes, John Heilemann, John House, Amy Hyams, Amy Johns, Steven Johnson, the incomparable Dana Jones, Kristine Kern, Toby Klayman and Joe Branchcomb, John Markoff and Leslie Terzian, Lesley McBride, Cindy and Mick McCaffrey, Rich Miller, Jamey and Carole Moore, Nancy Murphy, Francis Pisani, Lesa Porché, Lee and Tim Race, Mark Seiden, Larry Smith, Cyndi Stivers, Kathleen Sullivan, Karen Wickre, Tanya Wilkinson, Stacy Williams and Stacie Harkness, Allee Willis and Durwood Zedd.

I truly couldn't have finished — and might not have started in the first place — without them.

Financial support for The Hybrid Vigor Institute's xenotransplantation project came from the National Science Foundation, SES-0350493.

Hybrid Vigor's publication of *Intervention* has been supported to date by donations from Pandefense, the Tides Foundation, the Blue Mountain Center and many generous individuals, including Harriet Barlow, David Baron and Chris Bubser, Anita and Dana Cadonau-Huseby, Maggi Cary, Chris Desser and Kirk Marckwald, Katherine Fulton, Lisa Gansky, Don Hazen, Martin King, Rich Miller, Nancy Murphy, Norman Pearlstine, Tina Ragozzino and Stacy Williams.

BIBLIOGRAPHY
SELECTED FURTHER READING

When I began researching *Intervention*, I discovered that in order to write with any confidence I would have to return to first principles in almost every field touched by the intersection of risk, public policy and biotechnology. Rather than include my synthesis of that background material in the main text, I've appended here the books and studies that I found the most accessible and/or useful for answering my more basic questions. They are grouped by subject, insofar as that was possible, and listed alphabetically by author.

HISTORY AND PHILOSOPHY OF SCIENCE AND MATHEMATICS

THE EMERGENCE OF PROBABILITY, Ian Hacking

A compelling study of early ideas about probability and induction (Cambridge University Press, 1975), *The Emergence of Probability* has been called a "*tour de force* of historical scholarship." Fifteen years later, Hacking followed *Emergence* with the more famous *The Taming of Chance* (Cambridge University Press, 1990), which expanded his musings on probability into the realm of scientific reasoning. Warning: both books are written primarily for an academic audience.

THE MEASURE OF REALITY: QUANTIFICATION IN WESTERN EUROPE, 1250-1600, Alfred W. Crosby

Crosby's fascinating book *The Measure of Reality* (Cambridge University Press, 1997) is a must-read for the math-challenged like me, who never quite understood the relationship between mathematics and science. Until an Italian mathematician brought back Hindu-Arabic numerals from a business trip to Africa,

Western scholars delivered their theories about the physical attributes of the universe, such as velocity, temperature, and acceleration, as observations, not as data. In the space of less than a century, these practices quickly tumbled to the new methods of measurement and calculation made possible by the Arabic system. The effect on scientific discovery was profound.

THE STRUCTURE OF SCIENTIFIC REVOLUTIONS, Thomas S. Kuhn

Thomas Kuhn's classic 1962 essay, *The Structure of Scientific Revolutions* (University of Chicago Press, 3rd edition, 1996), is an essential filter through which to view the DNA-centric argument of heredity. Kuhn's essay is about big changes in theory, the tectonic shifts that take place periodically in the ongoing march of scientific discovery. He shows that there have always been long periods of what he called "normal science," which is what we would call DNA-dominated biology, interspersed with big disjunctive periods of "revolutionary science."

THE STUDY OF RISK

AGAINST THE GODS: THE REMARKABLE STORY OF RISK, Peter Bernstein

A best seller by the economist Peter Bernstein, *Against the Gods* (John Wiley & Sons, 1996) is the most accessible and comprehensive primer I could find about the truly remarkable history of risk. It includes fascinating historical evidence of how the discovery of probability theory and the development of statistical science led to a revolution in thinking about the nature of the physical world. Mention the word "risk" in conversation, and someone will invariably recommend it to you.

NORMAL ACCIDENTS: LIVING WITH HIGH-RISK TECHNOLOGIES, Charles Perrow

The publication of *Normal Accidents* (Princeton University Press, 1994) was one of the early public signals of a shift away from the purely mathematical interpretation of probabilities, risk and analysis. Now a classic, the book was written by the Yale sociologist Charles Perrow, who was the only social scientist asked to join the official investigation of the largest nuclear accident in U.S. history, the President's Commission on the Accident at Three Mile Island. The logic behind this odd term, "normal accidents," is that given the characteristics of complex systems — including humans — some failures are inevitable. But most of these accidents, Perrow asserts, are avoidable.

RISK, John Adams

A delightful and enlightening book by the geographer John Adams at University College London, *Risk* (UCL Press, 1995) details, among other things, his controversial theory of the "risk thermostat," showing that each of us has a different tolerance for risk. And on the other end, he shows how those who conduct risk analyses have their own biases and judgments that color the results they present to the public about the risks they measure.

RISK SOCIETY: TOWARDS A NEW MODERNITY, Ulrich Beck

German sociologist Ulrich Beck's compelling extended essay about risk and modernity, *Risk Society* (SAGE Publications Ltd., 1992), is a classic in the social science literature and a must-read for anyone interested in the cultural context in which risk exists. His theme, simply stated, is that modern science and technology have created a "risk society" in which the production of wealth has been overtaken by the production of risk. Scientists create the risks, he says, then fix them — a kind of unofficial "full employment act" for the technorati.

RISK, UNCERTAINTY, AND RATIONAL ACTION, Carlo C. Jaeger, Ortwin Renn, Eugene A. Rosa and Thomas Webler

Clearly and powerfully written for a mostly non-technical reader, the purpose of *Risk, Uncertainty, and Rational Action* (Earthscan, 2001) is to demonstrate how people think about risks and uncertainties, and particularly to dismantle the long-standing argument that people make decisions about risk wholly on rational thought. Encompassing the recommendations of the *Understanding Risk* study (details below), it moves beyond them to consider more specifically how stakeholders might best be taken into consideration in risk decisions.

RISK COMMUNICATION: A MENTAL MODELS APPROACH, M. Granger Morgan, Baruch Fischhoff, Ann Bostrom, Cynthia J. Atman

Written by four highly regarded risk and decision experts, *Risk Communication* (Cambridge University Press, 2002) is designed for a non-technical reader in order to make the "mental models" approach feasible for anyone who wants to try it. Illustrated with many real-world examples of risk communications on topics ranging from HIV/AIDS to health risks from electrical power fields, it also contains instructions on how to develop the kind of influence diagrams our team created for assessing the risks of xenotransplants.

BIOTECHNOLOGY AND GENETIC ENGINEERING

BIOHAZARD, Michael Rogers

Michael Rogers' highly readable account of "the most promising (and the most threatening) scientific research ever undertaken" is still the definitive chronicle of the early days after the invention of recombinant DNA. It includes Rogers' invaluable inside account of the famed Asilomar meeting in 1975. While no longer in print, *Biohazard* (Alfred A. Knopf, 1977) can still be found with relative ease online, as well as in used-book stores.

GENETICS & THE MANIPULATION OF LIFE: THE FORGOTTEN FACTOR OF CONTEXT, Craig Holdrege

Lynn Margulis, a distinguished university professor at the University of Massachusetts, Amherst, says that *Genetics & the*

Manipulation of Life (Lindesfarne Press, 1996) is "the single most accessible source not only of information (on cell biology) but of knowledge and wisdom." Holdrege, a high school biology teacher, writes simply and clearly to provide example after vivid example of how intimately entwined are living plants and animals with their environment; i.e., the context in which they are born, live and die. (Apparently in the United Kingdom, the book was published under the title, *A Question of Genes: Understanding Life in Context* (Floris Books, 1996).)

NATURE AND *NURTURE: THE COMPLEX INTERPLAY OF GENETIC AND ENVIRONMENTAL INFLUENCES ON HUMAN DEVELOPMENT AND BEHAVIOR*, edited by Cynthia García Coll, Elaine L. Bearer and Richard M. Lerner

Some of the articles in *Nature* and *Nurture* (Lawrence Erlbaum Associates, 2004) are too technical for a general audience to comfortably read and absorb. The volume deserves inclusion in this list because it presents such an array of compelling research results, from a wide variety of academic disciplines, to refute the "DNA determinism" argument.

RISK ASSESSMENT AND POLICY

IMPROVING RISK COMMUNICATION, Committee on Risk Perception and Communication

One of the four seminal studies on improving how society deals with the risks of science and technology, *Improving Risk Communication* (National Research Council, National Academy of Sciences, 1989) contains an appendix that's worth the price of the entire book. Called "Risk: A Guide to Controversy," and written for a non-technical audience by the risk expert Baruch Fischhoff of Carnegie Mellon University, this practical guide demonstrates how to characterize risk controversies along five essential dimensions, including "What are the limits to scientific estimates

mates of riskiness?" and "What are the (psychological) obstacles to laypeople's understanding of risks?"

SCIENCE & JUDGMENT IN RISK ASSESSMENT, Committee on Risk Assessment of Hazardous Air Pollutants

Another of the seminal risk studies from the National Academies, *Science & Judgment* (National Research Council, National Academy of Sciences, 1994) was the predecessor to *Understanding Risk*. As with the other risk studies I've mentioned, you'd be hard pressed to see many of its recommendations in action today. Yet its suggestions on how to use uncertainty as a tool in risk analysis, rather than avoiding it, can be generalized to all kinds of scientific and technological risks. The report is worth the price just to read the committee's eye-opening critique of what its client (in this case, the EPA) was doing wrong — and what it should do instead.

SCIENCE AND TECHNOLOGY ADVICE FOR CONGRESS, M. Granger Morgan and John M. Pena, Editors

Science and Technology Advice for Congress (RFF Press, 2003) is a compilation of expert assessments about how best to provide advice to government since the demise of the U.S. Office of Technology Assessment. Replete with examples and suggestions, it includes essays by some of today's most thoughtful social scientists working in this important area of study.

UNDERSTANDING RISK: INFORMING DECISIONS IN A DEMOCRATIC SOCIETY, Committee on Risk Characterization

Published in 1996 by the National Research Council of the National Academies and written for a general audience, this remarkable report was one of the first to seriously address, by way of methodology, the *scientific* importance of opening up the process of risk analysis to a wider group of people who are both interested in and affected by the issues.

Understanding Risk seems almost anarchic in comparison to the predictable days in the mid-20th century, when probabilistic risk analysis was on its own merits believable, acceptable, and in any case, simply the Way It Was Done. It throws down the gauntlet and says that risk analysis is a political, ethical and values-laden activity, period, and that it should be conducted with full participation by the people whose fate is at stake. "The aim is to describe a potentially hazardous situation in as accurate, thorough and decision-relevant manner as possible," the authors write, "addressing significant concerns of interested and affected parties, understandably and accessibly."

The study provides much useful detail about the "analytic-deliberative process," defining analysis as using "rigorous, replicable methods, evaluated under agreed protocols of an expert community — such as those of disciplines in the natural, social, or decision sciences, as well as mathematics, logic and law — to arrive at answers to factual questions." Deliberation is "any formal or informal process for communication and collective consideration of issues."

In addition to its exploration of the analytic-deliberative approach to risk, *Understanding Risk* details many useful examples of where the approach was used and the results, not all of which were positive. They are enlightening on their own merits.

BIBLIOGRAPHY

SOURCES AND CITATIONS (BY CHAPTER)

FRONT MATTER

Definition of intervention adapted from the 2003 American Society of Addiction Medicine, University. of Pennsylvania Health System, 2003, http://www.uphs.upenn.edu/addiction/berman/glossary/, accessed October 2006.

NOTE TO READERS

[1] Crosby, Alfred, *The Columbian Exchange: Biological and Cultural Consequences of 1492* (Greenwood Press, 1972), p. xiv-xv.

INTRODUCTION. (GENE)SIS

[1] I have served as a member of the Molecular Sciences Institute Board of Trustees since 2001. Brent served on the Hybrid Vigor Institute's board of directors from 2001 to 2003, and remains a member of the Institute's advisory board.

[2] Jaeger, Carlo, Ortwin Renn, Eugene A Rosa, Thomas Webler, *Risk, Uncertainty and Rational Action* (Earthscan, 2001).

[3] Preston, Richard, "The Genome Warrior," June 12, 2000, *The New Yorker*: 66-83.

[4] Kilman, Scott. "Use of Genetically Modified Seeds by U.S. Farmers Increases 18%," *The Wall Street Journal*, July 2, 2001.

CHAPTER 1. WHAT IF THE EXPERTS ARE WRONG?

[1] "Short Historical Timeline of Genome Sequencing," Argos Biotech, http://www.argosbiotech.de/700/omics/genomics/genometer.htm

[2] "President Clinton Announces the Completion of the First Survey of the Entire Human Genome," June 25, 2000, White House Press Release via Human Genome Project Information site, http://www.ornl.gov/sci/techresources/Human_Genome/project/clinton1.shtml, accessed October 2006.

[3] Saul, John Ralston, *Voltaire's Bastards: The Dictatorship of Reason in the West*, Penguin Books (1992), p 307

[4] Colborn, Theo, Dianne Dumanoski, John Peterson Myers, *Our Stolen Future*(Plume, 1997), p. 54

[5] Ibid., p. 65

[6] Ibid., p 57-58.

[7] "Office of Pollution Prevention and Toxics: PFOA and Fluorinated Telomeres," U.S. Environmental Protection Agency, available from http://www.epa.gov/opptintr/pfoa/index.htm, accessed October 2006.

[8] Hunt PA, KE Koehler, M Susiarjo, CA Hodges, A Ilagan, RC Voigt, S Thomas, BF Thomas, Hassold TJ. "Bisphenol A exposure causes meiotic aneuploidy in the female mouse," *Current Biology* 2003; 13(7):546-53.

[9] "Questions and Answers about BPA," from Bisphenol A, available from http://www.bisphenol-a.org/sixty-minutes2a.html, a website of American Plastics Council, http://www.plastics.org/s_plastics/doc_aboutus.asp?TRACKID=&CID=188&DID=403, accessed October 2006.

[10] "ToxFAQ for PBDEs," Agency for Toxic Substances and Disease Registry, U.S. Centers for Disease Control and Prevention, available from http://www.atsdr.cdc.gov/tfacts68-pbde.html, accessed October 2006.

[11] "PBDE SNUR Questions and Answers," Office of Pollution Prevention and Toxics, U.S. Environmental Protection Agency, available from http://www.epa.gov/oppt/pbde/pubs/qanda.htm, accessed October 2006.

[12] Iovine NM, Blaser MJ. "Antibiotics in animal feed and spread of resistant *Campylobacter* from poultry to humans," *Emerging Infectious Diseases*, Volume 10, No. 6, June 2004, available from: http://www.cdc.gov/ncidod/EID/vol10no6/04-0403.htm, accessed October 2006.

[13] "Initial Decision On Proposed Withdrawal of Baytril Poultry Nada," CVM Update, March 18, 2005, U.S. Food and Drug Administration Center for Veterinary Medicine, Office of Management and Communications, HFV-12, available from http://www.fda.gov/cvm/CVM_Updates/ baytri-lup.htm, accessed October 2006.

[14] "Enro/Cipro Art. No. R3111," R-Biopharm AG: Residues - Enro/Cipro, Food and Feed Analysis, available from http://www.r-biopharm.com/ foodandfeed/ridascreen_enro_cipro.php, accessed October 2006.

[15] "Stanley B. Prusiner, Autobiography" available from http://nobelprize.org/nobel_prizes/medicine/laureates/1997/prusiner-autobio.html, accessed October 2006.

[16] "Prion Diseases," from Chapter 8, *Principles of Molecular Virology* (Academic Press Inc., 2005), available from Microbiology @ Leicester, http://www-micro.msb.le.ac.uk/3035/prions.html, accessed October 2006.

[17] Angers, Rachel C., Shawn R. Browning, Tanya S. Seward, Christina J. Sigurdson, Michael W. Miller, Edward A. Hoover, Glenn C. Telling, "Prions in Skeletal Muscles of Deer with Chronic Wasting Disease," *Science*, published online January 26, 2006.

[18] Prusiner, ibid.

[19] Meinesz, Alexandre, excerpt from *Killer Algae* (University of Chicago Press, 1999), translated by Daniel Simberloff, available from http://www.press.uchicago.edu/Misc/Chicago/519228.html, accessed October 2006.

[20] Other than a science fiction writer, of course — in 1904, H.G. Wells published a novel about an invasive chemical foodstuff called *Herakleophorbia*. "The Food of the Gods and How It Came to Earth," available at Wikipedia, http://en.wikipedia.org/wiki/The_Food_of_the_Gods_and_How_It_Came_to_Earth, accessed October 2006.

[21] O'Hara, P, "The illegal introduction of rabbit haemorrhagic disease virus in New Zealand," *Rev. sci. tech. Off. int. Epiz.*, 2006, 25 (1), 119-123

[22] Hayes, RA and BJ Richardson, "Biological control of the rabbit in Australia: lessons not learned?" *TRENDS in Microbiology*, Vol.9 No. 9, September 2001, p. 459-460.

[23] "CSIRO Cane Toad Research," Publication - General, Commonwealth Scientific & Industrial Research Organisation, available from http://www.csiro.au/csiro/content/file/pfer,,.html, accessed October 2006.

[24] Nowak, Rachel, "Killer Virus," 10 January 01, *New Scientist,* available from http://www.newscientist.com/news/news.jsp?id=ns9999311, accessed October 2006.

[25] "Vaccine-evading mousepox virus ignites debate," CNN, Oct 31, 2003, via The Associated Press, available at http://www.cnn.com/2003/HEALTH/10/31/bioengineered.mousepox.ap/, accessed October 2006.

[26] Carlsen, William, "New documents show the monkey virus is present in more recent polio vaccine," July 22, 2001, *San Francisco Chronicle,* available at http://www.sfgate.com/cgi-bin/article.cgi?file=/chronicle/archive/2001/07/22/MN173141.DTL, accessed October 2006.

[27] "W. House Guts Global Warming Study," June 19, 2003, CBS News & The Associated Press, available at http://www.cbsnews.com/stories/200307/24/politics/printable564873.shtml, accessed October 2006.

[28] Vedantam, Shankar, "New EPA Mercury Rule Omits Conflicting Data: Study Called Stricter Limits Cost-Effective," Tuesday, March 22, 2005,*Washington Post*, Page A01, available from http://vedantam.com/epa-harvard-03-2005.html, accessed October 2006.

[29] Stein, Rob, "U.S. Says It Will Contest WHO Plan to Fight Obesity," January 16, 2004, *The Washington Post*, Page A03.

[30] "Special Interest Takeover," from summary, Citizens For Sensible Safeguards, available from http://www.sensiblesafeguards.org/sit.phtml, accessed October 2006.

CHAPTER 2. OF MICE, MEN AND UNCERTAINTY

[1] Tilghman, Shirley, "Biology in the Era of Complete Genomes," Watson Lecture, April 14, 2003, From Double Helix to Human Sequence — and Beyond, April 2003 Scientific Symposium, "50 Years of DNA Celebration," National Institutes of Health, available from http://genome.gov/Pages/News/webcasts/nhgri041403/WatsonLecture.ram.

[2] "25 Nobel Prize Winners in Support of Agricultural Biotechnology," AgBioWorld Foundation, available from http://www.agbioworld.org/declaration/nobelwinners.html. accessed October 2006.

[3] In 1943, American Oswald Avery proved that DNA carries genetic information. He even suggested DNA might actually be "the gene."

[4] Holdrege, Craig. *Genetics and the Manipulation of Life: The Forgotten Factor of Context*; (Lindisfarne Press, 1996), p 54-55.

[5] Holdrege p 54

[6] Rogers, Michael, *Biohazard* (John Wiley & Sons, 1977), p 7-8.

[7] Watson, James and John Tooze, *The DNA Story* (W H Freeman and Co., 1981), p vii.

[8] Mallet, J. “Hybridization as an invasion of the genome,” *TRENDS in Ecology and Evolution*, Vol.20 No.5, May 2005 p 229-237.

[9] Arnold, Michael L. and Edward J. Larson, “Evolution’s New Look,” *Wilson Quarterly*, August 2004, p. 60.

[10] Ibid.

[11] For an overview of this entire field, *Nature and Nurture: the Complex Interplay of Genetic and Environmental Influences on Human Behavior and Development*, edited by Cynthia Coll and Elaine Bearer of Brown University and Richard Lerner, Tufts University (Lawrence Erlbaum Associates, 2004).

[12] Weaver, I. N Cervoni, F Champagne, A D’alessio, S Sharma, J Secki, S Dymov, M Meaney. “Epigenetic Programming By Maternal Behavior,” *Nature Neuroscience*, Vol 7 No 8 Aug 2004 p 847.

[13] All quotes from interview with author, July 2004.

[14] Berg, Paul, “Asilomar and Recombinant DNA,” from Articles - Chemistry at Nobelprize.org, available at http://nobelprize.org/medicine/ articles/berg/, accessed October 2006.

CHAPTER 3. THE EFFECTS OF BIOTECH AT SCALE

[1] Fentiman, Smith and Meredith, “What are the Health Effects of Ionizing Radiation?”, Ohio State University Extension Research, RER-24, from http://extension.osu.edu/~rer/rerhtml/rer_24.html, accd. October 2006.

[2] “Global Status of Commercialized Biotech/GM Crops: 2004,” *ISAAA Executive Summary Preview*,. International Service for the Acquisition of Agri-biotech Applications, No. 32 - 2004.

[3] Ibid.

[4] *Gone to Seed: Transgenic Contaminants in the Traditional Seed Supply*, Union of Concerned Scientists, June 2004, http://ucsusa.org/food_and_environment/genetic_engineering/gone-to-seed.html, accessed Oct.2006.

[5] Barboza, David, “Illegal Rice Found Again In China’s Food Supply,” June 14, 2005, *The New York Times*, available at http://www.nytimes.com/2005/06/14/business/worldbusiness/14rice.html, accessed Oct. 2006.

[6] Hansen, Michael, “Problems with GM Papaya: Potential Human Health Effects of GE/GM Papaya,” May 2, 2005, Consumers Union of the United States.

[7] "Hawaiians warn against GM papaya," Bangkok Post, July 4, 2003, available at http://www.aseanbiotechnology.info/Abstract/23003866.pdf, accessed October 2006.

[8] Inbaraj, Sonny, "Thailand: Row Over GM Papaya to Surface at Environmental Meet," November 15, 2004, Inter Press Service News Agency, http://www.ipsnews.net/new_nota.asp?idnews=26273, accessed April 15, 2005.

[9] Bhattacharya, Shaoni, "Genetic engineers decaffeinate coffee," New Scientist.com News Service, June 18, 2003, available at http://www.newscientist.com/channel/life/gm-food/dn3851, accessed October 2006.

[10] Pollack, Andrew, "Genes from engineered grass spread for miles, study finds," September 21, 2004,*The New York Times*, A1.

[11] List from Center for Plant Responses to Environmental Stresses, Iowa State University Plant Sciences Institute, available at http://www.plantstress.iastate.edu/, accessed October 2006.

[12] Pearce, Fred, "China's GM Trees Get Lost In Bureaucracy," September 20, 2004, *New Scientist* Print Edition, available at http://www.newscientist.com/article.ns?id=dn6402, accessed October 2006.

[13] Rugh C, J, Senecff R Meagher and S Merkle., "Development of transgenic yellow poplar for mercury phytoremediation," *Nature Biotechnology* 1998, 16, 925-6; Bizily S, C Rugh and R Meagher, "Phytoremediation of hazardous organomercurials by genetically engineered plants," *Nature Biotechnology* 2000, 18, 213-5; Kramer U and A Chardonnens, "The use of transgenic plants in the bioremediation of soils contaminated with trace elements," *Applied Microbiology Biotechnology*, 2001, 55,661-72; Pilon-Smits E and M Pilon, "Breeding mercury-breathing plants for environmental cleanup," *Trends in Plant Science*, 2000, 5,235-6; all cited in "GM Trees Alert," *Science in Society*, No. 16, August 2002, Institute of Science in Society, available at http://www.i-sis.org.uk/full/GMtreesFull.php, accessed October 2006.

[14] "It's Your Health: Mercury and Human Health," Health Canada website, available at http://www.hc-sc.gc.ca/iyh-vsv/environ/merc_e.html, accessed October 2006.

[15] Wright, Karen, "Our Preferred Poison," *Discover,* Vol. 26 No. 03, March 2005, , available at http://www.discover.com/issues/mar-05/features/our-preferred-poison/, accessed October 2006.

[16] Ibid.

[17] "Biotech fights pollution with one tree at a time," CNN Science & Space, July 5, 2005, available at http://www.forestrycenter.org/headlines.cfm?RefID=75441 (original article no longer at CNN.com), accessed October 2006.

[18] Bhattacharya, Shaoni "Genetically-modified virus explodes cancer cells," June 1, 2004, *New Scientist* 13:39, available at http://www.newscientist.com/article.ns?id=dn5056, accessed October 2006.

[19] *Biological Confinement Of Genetically Engineered Organisms* (National Academy Press, 2004), p 161.

[20] Zandonella, Catherine, "GM bacteria may banish tooth decay," February 18, 2002, *New Scientist* 09:40, , available at http://www.newscientist.com/article.ns?id=dn1941, accessed October 2006.

[21] Cohen, Philip, " 'Living condom' could block HIV," September 9, 2003, *New Scientist* 10:12, available at http://www.newscientist.com/article.ns?id=dn4141, accessed October 2006.

[22] Zimmer, Carl, "Tinker, Tailor: Can Venter Stitch Together A Genome From Scratch?", February 14, 2003, *Science*, Vol. 299. no. 5609, pp. 1006 - 1007.

[23] Lederberg, Joshua, "The Microbial World Wide Web," April 14, 2000, *Science*, Vol 288, Issue 5464, p 291.

[24] Fox, Maggie, "Gene therapy did cause cancer in boys, study shows," October 16, 2003, Reuters Health Information, available via Cancerpage.com, http://www.cancerpage.com/news/article.asp?id=6402, accessed October 2006.

[25] "Gene Therapy," Human Genome Project Information, available at http://www.ornl.gov/sci/techresources/Human_Genome/medicine/gene therapy.shtml, accessed October 2006.

[26] U.S. Supreme Court, Diamond v. Chakrabarty, 447 U.S. 303 (1980), available at FindLaw, http://caselaw.lp.findlaw.com/scripts/getcase.pl?court=US&vol=447&inv ol=303, accessed October 2006.

[27] Engineering Life," interview with Ananda Mohan Chakrabarty, *The Hindu*, April 12, 2002, available at http://www.thehindu.com/thehindu/mp/2002/12/04/stories/20021204 00310200.htm, accessed October 2006.

CHAPTER 4. THE RISKS OF GOING NATIVE

[1] *Animal Biotechnology: Science-Based Concerns,* National Academy Press, 2002, p 4.

[2] Ibid., p. 11

[3] Coleman, Ron. E-mail to author, December 16, 2004.

[4] "How to add a fish gene to a tomato," BBC, October 30, 2002. http://www.bbc.co.uk/ science/genes/gm_genie/gm_science/index.shtml via the Wayback Machine, accessed October 2006.

[5] Coleman, Ron. E-mail to author, December 16, 2004.

[6] "Bugs In The System: Issues in the Science and Regulation of Genetically Modified Insects," Pew Initiative on Food and Biotechnology, available at http://pewagbiotech.org/research/bugs/, accessed October 2006.

[7] 'Transgenic Mosquitoes Are Less Fertile Than Their Counterparts In Nature,' University of California, Riverside, press release, January 13, 2004; Irvin N, Hoddle MS, O'Brochta DA, Carey B, Atrkinson PW, "Assessing fitness costs for transgenic *Aedes aegypti* expressing the GFP marker and transposase genes," Proceedings of the National Academy of Sciences, published online before print, January 7, 2004, www.pnas.org/cgi/doi/10.1073/ pnas.0305511101

[8] "Scientists plan to wipe out malaria with GM mosquitoes," The Guardian, Sept 3, 2001. http://www.guardian.co.uk/print/ 0,3858,4249413-103690,00.html

[9] Greenpeace GM contamination register, http://www.gmcontaminationregister.org/, accessed June 13, 2005.

[10] *Biological Confinement of Genetically Engineered Organisms,* National Academy Press, 2004.

[11] Ellstrand, Norman C. "Going to 'Great Lengths' to Prevent the Escape of Genes that Produce Specialty Chemicals," *Plant Physiology,* August 2003, Vol. 132, pp. 1770-1774, available at http://www.plantphysiol.org, accessed October 2006.

[12] Ibid.

[13] Ibid.

[14] Hileman, Bette, "Prodigene and StarLink Incidents Provide Ammunition to Critics," *Chemical & Engineering News*, June 9, 2003, Volume 81, No.

23, p 25-33, available at http://pubs.acs.org/cen/coverstory/8123/8123biotechnologyb.html, accessed October 2006.

CHAPTER 5. WHAT GETS MEASURED IS WHAT MATTERS

[1] All Thurman material from conversations or e-mails with author, over the course of several months in 2003.

[2] Adams, John, *Risk,* UCL Press Ltd., 1995

[3] Karp, Hal, "Kids at Risk: when seat belts are NOT enough," November 1999, *Reader's Digest*, U.S. edition, as cited in "Seat Belt Laws: Why You Should Be Worried," Galway Cycling Campaign, Nov 2000.

[4] Paulos, John Allen, *Innumeracy: Mathematical Illiteracy and Its Consequences (Hill and Wang, 2001), p 5*

[5] *Chemical and Engineering News*, 1979.

[6] Fischhoff, B, and I. Fischhoff, "Publics Opinions about Biotechnology," *AgBioForum,* Volume 4, Number 3&4, 2001, Pages 155-162

[7] "GM Nation? About the debate," GM Nation: The Public Debate, available from http://www.gmnation.org.uk/ut_09/ut_9_6.htm, accessed October 2006.

[8] Ibid.

[9] Ibid.

[10] Brown, Paul, "GM Crops to get go-ahead: Leaked papers reveal decision," February 19, 2004, *The Guardian,* Page 1.

[11] Robert Kohler, *Lords of the Fly: Drosophila Genetics and the Experimental Life.* (Chicago University Press, 1994)

[12] Conversation with author.

[13] Ibid.

[14] Fischhoff, Baruch, "Scientific Management of Science?," *Policy Sciences,* Vol. 33, p 73-87, 2000.

[15] Interview with author, September 9, 2004.

[16] Niederman and Boyum, *What the Numbers Say: A Field Guide to Mastering Our Numerical World* (Broadway Books, 2003), p 60

[17] Ibid., 64

[18] Kolata, Gina, "It Was Medical Gospel, But It Wasn't True," May 30, 2004, *The New York Times.*

CHAPTER 6. POLITICS, SCIENCE AND SUBSTANTIAL EQUIVALENCE

[1] "The Flavr Savr Arrives," copy of original U.S. Food and Drug Administration press release, dated May 18, 1994,at Access Excellence: the National Health Museum, About Biotech, available at http://www.accessexcellence.org/RC/AB/BA/Flavr_Savr_Arrives.html, accessed October 2006.

[2] "Biotechnology of Food," U.S. Food and Drug Administration, Center for Food Safety and Applied Nutrition, May 18, 1994, available at http://www.cfsan.fda.gov/~lrd/biotechn.html, accessed October 2006.

[3] Food and Agriculture Organization/World Health Organization, *Biotechnology and Food Safety, Report of a Joint FAO/WHO Consultation*, Rome, October 1996. FAO Food and Nutrition Paper 61, cited in Levidow and Murphy, *The Decline of Substantial Equivalence: How Civil Society demoted a Risky Concept* 5, Paper for Conf. at Institute of Development Studies, Dec. 12-13, 2002 ("Science and citizenship in a global context: challenges from new technologies").

[4] *Federal Register,* Vol. 54, No. 104 (1992), p. 22991, from "Should Genetically Modified Foods be Labeled," by Craig Holdrege, http://www.saynotogmos.org/regulatory_2.htm

[5] Federal Register vol. 59, No. 98 (1994), pp. 26700-26711, from "Should Genetically Modified Foods be Labeled," by Craig Holdrege, http://www.saynotogmos.org/regulatory_2.htm

[6] "Padgette, S. et al. "The Composition of Glyphosate-Tolerant Soybean Seeds Is Equivalent to That of Conventional Soybeans", *Journal of Nutrition*, Vol. 126: 702-716, 1996, cited in "Safety Assessment of Roundup Ready Soybean Event 40-3-2," Monsanto Corporation, available at www.monsantoinfo.dk/nyhedsbrev/Summary.pdf, accessed October 2006.

[7] "Bertram Rowland and the Cohen/Boyer Cloning Patent," George Washington University Law School, IP & Technology Law, available at http://www.law.gwu.edu/Academics/Academic+Focus+Areas/IP+and+Technology+Law/Alumni+Patents/Bertram+Rowland+and+the+Cohen+Boyer+Cloning+Patent.htm, accessed October 2006.

[8] SEMATECH: History," SEMATECH, available at http://sematech.org/corporate/history.htm, accessed October 2006.

[9] Eichenwald, Kurt, Gina Kolata and Melody Petersen, "Biotechnology Food: From the Lab to a Debacle," January 25, 2001, *The New York Times*, available at http://www.nytimes.com/2001/01/25/business/25FOOD.html?pagewanted=all&ei=5070&en=3e73ade60d327b2a&ex=114153480, accessed October 2006.

[10] "Statement by Principal Deputy Press Secretary Speakes on the Coordinated Framework for the Regulation of Biotechnology," June 20, 1986, Archives of the Ronald Reagan Presidential Library, available at http://www.reagan.utexas.edu/archives/speeches/1986/62086a.htm, accessed October 2006.

[11] *Field Testing Genetically Modified Organisms: Framework For Decisions*, NRC 1989, executive summary, http://books.nap.edu/catalog/1431.html. Committee list is at: http://www.nap.edu/catalog/blurb_catalog.phtml?val1=1431&val2=comlist

[12] "Index: Key FDA Documents Revealing Hazards Of Genetically Engineered Foods," Alliance for Bio-Integrity, available at http://www.biointegrity.org/FDAdocs, accessed October 2006.

[13] "Landmark Lawsuit Challenges FDA Policy on Genetically Engineered Food," Alliance for Bio-Integrity Press Release, available at http://www.biointegrity.org/Lawsuit.html, accessed October 2006.

[14] Kahl, Linda to James Maryanski, January 8, 1992, internal FDA memo,Alliance for Bio-Integrity, available at http://www.biointegrity.org/FDAdocs/01/01.pdf, accessed October 2006.

[15] Memorandum from Dr Fred Hines to Dr Linda Kahl, June 16, 1993, "FLAVR SAVR Tomato: Pathology Branch's Evaluation of Rats with Stomach Lesions From Three Four-Week Oral (Gavage) Toxicity Studies," Alliance for Bio-Integrity, available at http://www.biointegrity.org/FDAdocs/17/view1.html, accessed October 2006.

[16] "Memorandum from Dr Fred Hines to Dr Linda Kahl, December 10, 1993, "FLAVR SAVR Tomato: Pathology Branch's Remarks to Calgene Inc.'s Response to FDA Letter of June 29, 1993," available at http://biointegrity.org/FDAdocs/18/view1.html, accessed October 2006.

[17] "FDA'S Policy for Foods Developed by Biotechnology," CFSAN Handout 1995, from a chapter in the proceedings of American Chemical Society Symposium Series No. 605, 1995, presented by J. H. Maryanski, Strategic Manager for Biotechnology, U.S. Food and Drug Administration, Center for Food Safety and Applied Nutrition, available at http://www.cfsan.fda.gov/~lrd/biopolcy.html, accessed October 2006.

[18] Kasler, Dale and Edie Lau, "At Calgene, a harvest of uncertainty," May

7, 2000, *Sacramento Bee*, , available at http://dwb.sacbee.com/static/archive/news/projects/biotechnology/, accessed October 2006.

[19] "What happened to the Flavr Savr," *Chemical & Engineering News*, April 19, 1999, Vol. 77, No. 16, available from http://pubs.acs.org/hotartcl/cenear/990419/7716bus1box3.html, accessed October 2006.

[20] "Modern plant breeding v. traditional plant breeding," FDA Consumer Magazine, November -December 2003 issue, available at http://fda.gov/fdac/features/2003/ld-plantDNA.html, accessed October 2006.

[21] Genewatch Crop Line Datasheet, http://www.genewatach.org/GeneSrch/_scripts/LineData.asp. Crop line is GTS 40-3-2.

[22] "Elements of Precaution: Recommendations for the Regulation of Food Biotechnology in Canada," expert panel report on future of food biotechnology by the Royal Society of Canada for Health Canada Canadian Food Inspection Agency and Environment Canada. January 2001, available at http://www.rsc.ca//files/ publications/expert_panels/foodbiotechnology/GmreportEN.pdf, accessed October 2006.

[23] Nordlee, Julie A., M.S., Steve L. Taylor, Jeffrey A. Townsend, Laurie A. Thomas and Robert K. Bush, "Identification of a Brazil-Nut Allergen in Transgenic Soybeans," March 14, 1996, *New England Journal of Medicine,* Volume 334:688-692.

[24] "What do Experts say About the Potential Human Health Effects of Genetically Engineered Corn?", white paper, Friends of the Earth, August 2003, available from http://www.humboldt.org.ni/transgenicos/docs/what_experts_says_human_effects.pdf, accessed October 2006.

[25] "Reports of Starlink's Health Effects Were Greatly Exaggerated," June 2001, Center for Consumer Freedom, available at http://www.consumerfreedom.com/news_detail.cfm/headline/913, accessed October 2006.

[26] "Unlikely Reactions: Identifying Allergies to GM Foods," Pew Initiative on Food and Biotechnology, 2004, available at http://pewagbiotech.org/buzz/display.php3?StoryID=12, accessed October 2006.

[27] Young, Emma, "GM pea causes allergic damage in mice," *New Scientist* 11:18, 21 November 2005, available at http://www.newscientist.com/article.ns?id=dn8347, accessed October 2006.

[28] "Feeding trials to test GM food safety," November 26, 2005, Government of Western Australia Media Statement, available at http://www.mediastatements.wa.gov.au/media/media.nsf/news/45d1988205b503a4482570c7000e0793?opendocument, accessed October 2006.

[29] McGinnis, Mike, "Beef industry torn over Creekstone issue," April 20, 2004, *High Plains Journal*, available at http://www.hpj.com/archives/2004/apr04/BeefindustrytornoverCreekst.CFM, accessed October 2006; "Creekstone Farms Premium Beef Files Lawsuit Challenging USDA's Ban on Voluntary BSE Testing," March 2006, available at http://www.creekstonefarmspremiumbeef.com/news_bse_press.html, accessed October 2006.

[30] *Safety of Genetically Engineered Foods: Approaches to Assessing Unintended Health Effects* (National Academy Press, 2004).

[31] "Composition of Altered Food Products, Not Method Used to Create Them, Should Be Basis for Federal Safety Assessment," July 27, 2004, press release issued by The National Academies, available at http://www8.nationalacademies.org/onpinews/newsitem.aspx?RecordID=10977, accessed October 2006.

[32] *Safety of Genetically Engineered Foods: Approaches to Assessing Unintended Health Effects*, p 8-14.

CHAPTER 7. SILENT GENES

[1] Heinemann, JA, Sparrow, AD and Traavik, T. 2004. "Is confidence in the monitoring of GE foods justified?", *Trends in Biotechnology* 22, 331-336.

[2] "Suspected GM Sweet Corn: Questions and Answers," July 5, 2003, Press Release, Ministry of Agriculture and Forestry, *Scoop NZ Independent News*, http://www.scoop.co.nz/stories/SC0307/S00030.htm, accessed October 2006.

[3] "Japan ends U.S. long-grain rice imports," August 19, 2006, Associated Press via CBS News, available at http://www.cbsnews.com/stories/2006/08/20/ap/business/mainD8JJSCL00.shtml and http://news.yahoo.com/s/ap/japan_us_rice, accessed October 2006.

[4] Anklam, E. et al., "Analytical methods for detection and determination of genetically modified organisms in agricultural crops and plant-derived food products," *Eur. Food Res. Technol.* 2002, 214,3-26; and Matsuoka, T. et al. "Detection of recombinant DNA segments introduced to genetically modified maize (Zea mays)," *J. Agric. Food Chem.* 2002, 50, 2100-2109; both cited in Heinemann, et al. "Is confidence in monitoring of GE foods justified?" *Trends in Biotechnoogy,* 22, 331-336 (2004).

[5] Conversation with author.

[6] U.S.A. Patriot Act, Public Law 107-56—Oct. 26, 2001; Sec. 817,

Expansion of the Biological Weapons Statute.

[7] Golding, M. C., C. R. Long, M. A. Carmell, G. J. Hannon, M. E. Westhusin, "Suppression of prion protein in livestock by RNA interference," *Proceedings of the National Academy of Sciences*, published online Mar 27, 2006; Volume 103, p5285-5290; .

[8] According to Roger Brent, 3 to 8 plasmids that make RNAs can be purchased from Origene Corp. for $380, with greater than 95% chance that at least one will inactivate the action of that gene product.

[9] Collis SJ, Swartz MJ, Nelson WG, DeWeese TL, "Enhanced radiation and chemotherapy-mediated cell killing of human cancer cells by small inhibitory RNA silencing of DNA repair factors," *Cancer Research* 2003 Apr 1; 63(7): 1550-4.

[10] Lewis, Benjamin P, CB Burge, DP Bartel, "Conserved Seed Pairing, Often Flanked by Adenosines, Indicates that Thousands of Human Genes are MicroRNA Targets," Letter to Editor, *Cell*, Vol. 120, 15-20, January 14, 2005.

[11] Nelson, P, M Kiriakidou, A Sharma, E Maniataki, Z Mourelatos, "The microRNA world: small is mighty," *TRENDS in Biochemical Science*, Vol. 28, No. 10, October 2003, p 534

[12] Zwahlen C, Hilbeck A, Howald R, Nentwig W, "Effects of transgenic *Bt* corn litter on the earthworm Lumbricus terrestris," *Mol Ecol.* 2003 Apr;12(4):1077-86.

[13] Coghlan, Andy, "Pollution triggers bizarre behavior in animals," September 3, 2004, *New Scientist*, available at http://www.newscientist.com/article.ns?id=dn6343, accessed October 2006.

[14] Pryme, IF and Lembcke, R, "In Vivo Studies on Possible Health Consequences of Genetically Modified Food and Feed — With Particular Regard to Ingredients Consisting of Genetically Modified Plant Materials," *Nutrition and Health*, 2003, Vol. 17, pp. 1-8 0260-1060/03.

[15] "Food Safety and Substantial Equivalence," Biotechnology Industry Organization, available at http://www.bio.org/foodag/background /foodsafety.asp, accessed October 2006.

[16] Paabo, S, M Kohn, M Hoss, "Excrement analysis by PCR," *Nature*, Vol 359, 17 September 1992, p 199.

[17] Coghlan, Andy, "GM crop DNA found in human gut bugs," July 18, 2002, *New Scientist* News Service, available at http://newscientist.com/

article.ns?id=dn2565, accessed October 2006.

[18] Doerfler, Walter, *Foreign DNA in Mammalian Systems*, (Wiley-VCH Verlag GmbH., 2000), synopsis at http://www.powells.com/cgi-bin/biblio?inkey=4-3527300899-0, accessed October 2006.

[19] "International Documents and Scientific Publications on Plant Biotechnology and the Safety Assessment of Food Products Derived from Plant Biotechnology," ILSI International Food Biotechnology Committee, September 2004, available at http://orig.ilsi.org/file/ACF990D.pdf, accessed October 2006.

[20] Forsman, A., D. Ushameckis, et al. "Uptake of amplifiable fragments of retrotransposon DNA from the human alimentary tract," *Mol Genetics and Genomics* 2003 Dec; 270(4):362-8, published online October 11, 2003, available at Springer Link, http://www.springerlink.com/content/028yu0d3eja569da/, accessed October 2006.

CHAPTER 8. THE PROMISE OF TRANSGENICS

[1] Lambrecht, Bill. *Dinner At The New Gene Café* (St. Martin's Press, 2001), p. 139

[2] Ibid.

[3] "Reducing Poverty and Hunger: the Critical Role of Financing for Food, Agriculture and Rural Development," U.N. Food and Agriculture Organization, International Fund for Agricultural Development, and World Food Programme, March 2002, paper prepared for the International Conference on Financing for Development, Monterrey, Mexico, available at http://www.fao.org/docrep/003/Y6265e/y6265e00.htm, accessed October 2006.

[4] "Harvesting Poverty: The Unkept Promise," December 30, 2003, *The New York Times*, Editorials, abstract available at http://select.nytimes.com/gst/abstract.html?res=F30D15FF3B5A0C738FDDAB0994DB404482&showabstract=1, accessed October 2006.

[5] Benbrook, Charles M., "When Does It Pay to Plant Bt Corn? Farm-Level Economic Impacts of Bt Corn, 1996-2001," Benbrook Consulting Services, December 2001, available at http://www.biotech-info.net/Bt_farmlevel_IATP2001.html, accessed October 2006.

[6] Benbrook, Charles M. "Genetically Engineered Crops and Pesticide Use in the United States: The First Nine Years," Technical Paper Number 7,

BioTech InfoNet, October 2004, available at http://www.biotech-info.net/technicalpaper7.html, accessed October 2006.

[7] Ibid., p. 8

[8] Coordination of Delegations Department, Argentinian Agriculture, Livestock, Fishing & Food Secretariat (SAGPYA), Agricultural Statistics, accessed July 2005; http://www.sagpya.mecon.gov.ar/http-hsi/bases/oleagi.htm

[9] Branford, Sue, "Argentina's Bitter Harvest," April 17, 2004, *New Scientist*, Issue 2443, p 40-43, available to subscribers at http://www.newscientist.com/channel/life/gm-food/mg18224436.100-argentinas-bitter-harvest.html, accessed October 2006.

[10] Trigo, Eduardo, Daniel Chudnovsky, Eugenio Cap, Andrés López, "Genetically Modified Crops In Argentine Agriculture: An Open Ended Story," from the original *"Los transgénicos en la agricultura argentina. Una historia con final Abierto,* prepared by *Centro de Investigaciones para la Transformación* (CENIT), Libros del Zorzal, Buenos Aires, Argentina, 2002. http:// (www.fund-cenit.org.ar), accessed October 2006.

[11] Joensen L, Semino S, Paul H, *Argentina: A Case Study on the Impact of Genetically Engineered Soya*, p. 22. Report prepared for the Gaia Foundation, English summary available at http://gaiafoundation.org/resources/general.php?pub_id=356, accessed October 2006..

[12] "*New Scientist*, ibid.

[13] Faccini, Delma, Rosario National University, Eng., *AgroMensajes Magazine*, No. 4,4 p 5, December 2000; cited in *New Scientist*, ibid.

[14] Daniels, Roger, Caroline Boffey, Rebecca Mogg, Joanna Bond, Ralph Clarke, "The Potential For Dispersal Of Herbicide Tolerance Genes From Genetically-Modified, Herbicide-Tolerant Oilseed Rape Crops To Wild Relatives," final report to DEFRA, July 2005, available at http://www.defra.gov.uk/environment/gm/research/pdf/epg_1-5-151.pdf, accessed October 2006; Hopkin, Michael, "Transgenic crop may have bred with wild weed," *Nature*, published online July 25 ,2005, available at http://www.nature.com/news/2005/050725/full/050725-2.html, accessed October 2006.

[15] *New Scientist*, ibid.

[16] Reis, EM and Carmona, M, "Roya de la soja: Diagnóstico, epidemiología y manejo," *Revista Producción*, Jan. 4, 2004, http://produccion.com.ar/2004/04ene_12.htm; cited in Joensen L, et al., p. 18.

[17] Ibid, p 19.

[18] Hayes TB, A Collins, M Lee, M Mendoza, N Noriega, AA Stuart, A Vonk, "Hermaphroditic, demasculinized frogs after exposure to the herbicide atrazine at low ecologically relevant doses," *Proc. Natl. Acad. Sci.* USA, 99, 5476 (2002).

[19] *New Scientist*, ibid.

[20] Ibid.

[21] Boy, Adolfo, "Transgénicos, Fracaso del Modelo Agropecuario Argentina," 2003, cited in Joensen L, et al..

[22] Ibid, p15.

[23] *New Scientist*, ibid.

[24] National Cottonseed Products Assn, "Twenty Facts about Cottonseed Oil," available at http://www.cottonseed.com/publications/facts.asp, accessed October 2006.

[25] Qayum, Abdul and Sakkhari, Kiran, "Did *Bt* Cotton Fail A.P. Again In 2003-2004? A season long study of the performance of Bt Cotton in Andhra Pradesh, India," Deccan Development Society, AP Coalition in Defence of Diversity, Permaculture Association of India, available at http://www.eldis.org/static/DOC14627.htm, accessed October 2006.

[26] Ibid., p. 24

[27] "Performance of Bollgard Cotton in 2003," March 26, 2004, A.C. Nielsen and ORG Centre for Social Research.

[28] Jayaraman, KS, Jeffrey L. Fox, Hepeng Jia & Claudia Orellana, "Indian *Bt* gene monoculture, potential time bomb," *Nature Biotechnology* 23: 158., available to subscribers at http://www.nature.com/news/2005/050131/full/nbt0205-158.html, accessed October 2006.

[29] Kranthi, K.R. "Bollworm resistance to Bt cotton in India," Letter to the Editor, *Nature Biotechnology*, Volume 23, No. 12, 2005, p. 1476-77.

[30] S. Kranthi, K. R. Kranthi*, P. M. Siddhabhatti and V. R. Dhepe, "Baseline toxicity of Cry1Ac toxin against spotted bollworm, *Earias vittella* (Fab) using a diet-based bioassay," *Current Science* 87, No. 11, 10 December 2004, available at www.ias.ac.in/currsci/dec102004/1593.pdf, accessed October 2006.

[31] Mudur, G.S., "Govt spies genetic cotton faults," July 27, 2005, *The Telegraph*, Calcutta, available at http://www.telegraphindia.com/

1050727/asp/nation/story_5039613.asp#, accessed October 2006.

[32] Ibid.

CHAPTER 9. THE TRICKY CALCULUS OF COST AND BENEFIT

[1] Ravetz, Jerry. "Safety in the globalising knowledge economy: an analysis by paradoxes", *Journal of Hazardous Materials* 86 (2001) 1-16.

[2] Wright, Karen, "Our Preferred Poison," *Discover* Vol. 26 No. 03 | March 2005, available at http://www.discover.com/issues/mar-05/features/our-preferred-poison/, accessed October 2006.

[3] Vedantam, Shankar, "New EPA Mercury Rule Omits Conflicting Data: Study Called Stricter Limits Cost-Effective," March 22, 2005, *Washington Post*, available at http://vedantam.com/epa-harvard-03-2005.html, accessed October 2006.

[4] Ibid.

[5] Ibid.

[6] "Monsanto v. U.S. Farmers," 2004-2005, A Report by the Center for Food Safety, p 31, available at http://www.centerforfoodsafety.org/pubs/CFSMOnsantovsFarmerReport 1.13.05.pdf, accessed October 2006.

[7] "Saving Seed Is Latest Tech Piracy," by the Associated Press, January 14, 2005, as viewed at http://wired-vig.wired.com/news/technology/0,1282,66282,00.html

[8] "Understanding the WTO: Basics: The Uruguay Round,", World Trade Organization, available at http://www.wto.org/english/thewto_e/whatis_e/tif_e/fact5_e.htm, accessed October 2006.

[9] Gurdial Nijar Singh, July 1999, Interview at the Intersessional Meeting of the U.N. Convention on Biodiversity, as cited in McAfee, Kathleen, "Are Economic and Genetic Reductionism Linked in Biotechnology Science and Policy?" presentation at Society for the Social Study of Science Cambridge, MA, November 2001. More on reductionism can be found in "Neoliberalism on the molecular scale. Economic and genetic reductionism in biotechnology battles," *Geoforum* 34 (2003) 203-219, available at http://www.kmcafee.com/PDF/Molecular_Neoliberalism.pdf, accessed October 2006.

[10] All quotes this chapter from conversations with author.

[11] "Dioxins and their effects on human health," World Health Organization Fact Sheet No. 225, June 1999, available at http://www.who.int/mediacentre/factsheets/fs225/en/, accessed October 2006.

[12] All O'Brien's comments from "What Other Course Is Open To Us: Risk Analysis, Alternatives, and Democracy," presentation at December 2004 Society for Risk Analysis annual meeting, available at http://www.sra.org/events_2004_meeting.php, accessed October 2006.

[13] TIME Magazine, cited in "The False Hope Of Golden Rice," Greenpeace briefing, March 2005, available at http://www.greenpeace.org/raw/content/italy/ufficiostampa/rapporti/golden-rice-briefing-0305.pdf#search=%22"The%20False%20Hope%20Of%20Golden%20Rice%2C"%22, accessed October 2006

[14] Potrykus, I, "Golden Rice and beyond," *Plant Physiology* 2001, 125: 1157-1161, as quoted in Greenpeace, ibid.

[15] Coffman, R. McCouch, SM & Herdt, RW. "Potentials and Limitations of Biotechnology in Rice," United Nations Food and Agriculture Organization Rice Conference, Rome, Italy, February 12-13, 2004, http://www.fao.org/rice2004/en/pdf/coffman.pdf (accessed October 2006), as cited in Greenpeace, ibid.

[16] "Dietary reference intakes for vitamin A, vitamin K, arsenic, boron, chromium, copper, iodine, iron, manganese, molybdenum, nickel, silicon, vanadium, and zinc: A Report of the Panel on Micronutrients Subcommittees on Upper Reference Levels of Nutrients and of Interpretation and Uses of Dietary Reference Intakes, and the Standing Committee on the Scientific Evaluation of Dietary Reference Intakes, Food and Nutrition Board," U.S. Institute of Medicine (National Academies, 2002); Potrykus, I. "Extreme Precautionary Regulation Is *The* Obstacle For Public Goods Green Biotechnology," Presentation at REDBIO 2004, meeting of the Latin America/Caribbean Plant Biotechnology Network, June 21-25, 2004, Santo Domingo, Dominican Republic. Available at www.redbio.org/rdominicana/redbio2004rd/Memoria_REDBIO_2004/plenarias-PDF/p01-PDF/p01.pdf, accessed October 2006. Both as cited in Greenpeace, ibid.

[17] Datta, K., Baisakh,. N., Oliva, N., Torrizo, L., Abrigo, E., Tan, J., Rai, M., Rehana, S., Al-Babili, S., Beyer, P., Potrykus, I. & Datta, S.K., "Bioengineered 'golden' indica rice cultivars with ß -carotene metabolism in the endosperm with hygromycin and mannose selection systems," *Plant Biotechnology Journal* 2003, 1: 81-90, as cited in Greenpeace, ibid.

[18] Ye, X., Al-Babili, S., Klöti, A., Zhang, J., Lucca, P., Beyer, P. & Potrykus, I. "Engineering the provitamin A (ß -carotene) biosynthetic pathway into (carotenoid-free) rice endosperm," *Science*, 2000, 287: 303-305, and Kuiper, H.A., Kleter, G.A., Noteborn, H.P.J.M. & Kok, E.J. "Assessment of the food safety issues related to genetically modified foods," *Plant Journal* 2001, 27: 503-528; both cited in Greenpeace, ibid.

[19] Beyer, P., Al-Babili, S., Ye, X, Lucca, P., Schaub, P., Welsch, R. & Potrykus, I. "Golden Rice: introducing the ß -carotene biosynthesis pathway into rice endosperm by genetic engineering to defeat vitamin A deficiency," *Journal of Nutrition* 2002 132: 506S-510S, as cited in Greenpeace, ibid.

[20] Crolly, Hannelore. "Syngenta Halts Genetic Engineering Projects in Europe," November 29, 2004, *Die Welt,* available at http://www.gene.ch/genet/2004/Dec/msg00011.html, accessed October 2006.

[21] Zimmermann, R., M. Qaim, "Potential health benefits of Golden Rice: a Philippine case study," *Food Policy*, 2004, Vol. 29, No. 2, pp. 147-168.

[22] Micronutrient Initative & UNICEF, "Vitamin & Mineral Deficiency. A Global Progress Report," Micronutrient Initiative; Ottawa, Canada, available at http://www.micronutrient.org/reports/, accessed October 2006, as cited in Greenpeace, ibid.

[23] Lorch, A., "Vitamin A Deficiency: Diverse Causes, Diverse Solutions," for Greenpeace International, 2005, at http://greenpeaceweb.org/gmo/vitamina.pdf, accessed October 2006.

[24] Xiangzhong (Jerry) Yang, "Cloning and other reproductive technologies for applications in transgenics," presentation at National Institutes of Standards and Technology, Chemistry and Life Sciences Office, available at http://www.atp.nist.gov/atc/atc-22.htm, accessed October 2006.

[25] Margawati, Endang Tri. "Transgenic Animals: Their Benefits To Human Welfare," available at http://www.actionbioscience.org/biotech/margawati.html#articlereferences, accessed October 2006.

[26] Shreeve, Jamie, "The Other Stem Cell Debate," April 10, 2005, *New York Times Magazine*, p. 42.

[27] *Scientific and Medical Aspects of Human Reproductive Cloning* (National Academies, 2002), p. 41.

[28] Ibid., p 49

[29] Interview with author, 2005.

CHAPTER 10. OUR APPOINTED ARBITERS OF RISK

[1] Fischhoff, Baruch. "Cost-benefit analysis and the Art of Motorcycle Maintenance," *Policy Studies* 8 (1977), pp 177-202.

[2] All quotations from interviews with author.

[3] Caruso D & D Rhoten, "Lead, Follow, or Get Out of the Way: Sidestepping the Barriers to Effective Practice of Interdisciplinarity," 2001, The Hybrid Vigor Institute, http://www.hybridvigor.net/.

[4] Interview with author.

[5] Jensen, K, and Fiona Murray, "Intellectual Property Landscape of the Human Genome," *Science* 14 October 2005: Vol. 310. no. 5746, p 239 - 240, available at http://www.sciencemag.org:80/cgi/content/full/310/5746/239, accessed October 2006.

[6] Wilson, E. *Chem. Eng. News* **79**, 41–49 (2001)., cited in Peter Shorett, Paul Rabinow, and Paul R. Billings 'The changing norms of the life sciences,' *Nature Biotechnology*, February 2003, Vol. 21, p 123.

[7] Lawrence, Stacy. "Biotech patenting matures," *Nature Biotechnology*, Volume 22, Number 10, October 2004.

[8] Washburn, Jennifer, "The Kept University," *Atlantic Monthly*, March 1, 2000.

[9] Blumenstyk, Goldie, "Peer Review Slams Biotech Giant Takeover of University Biology Department," July 30, 2004, *Chronicle of Higher Education.*

[10] "Good 'citizenship' or good business?" *Nature Genetics* Volume 36 Number 10 Oct 2004, p 1025

[11] Willman, David, "New rules will cost dissidents at NIH," March 3, 2005, *Los Angeles Times*, available at http://www.latimes.com/news/nationworld/nation/la-na-nih3mar03,1,2122772.story?coll=la-headlines-nation, accessed October 2006.

[12] Willman, David, "Stealth merger: drug companies and government medical research; some of the National Institutes of Health's top scientists are also collecting paychecks and stock options from biomedical firms," Dec 7, 2003, *Los Angeles Times*, p A1.

[13] Pollan, Michael, "Playing God In the Garden," Oct. 25, 1998, *New York Times Magazine*, available at http://www.michaelpollan.com/article.php?id=73, accessed October 2006.

[14] "FDA/Healthcare Practice," from King & Spalding LLP website, available at http://www.kslaw.com/portal/server.pt?space=KSPublicRedirect&control=KSPublicRedirect&PracticeAreald=200, accessed October 2006.

[15] All information on individuals from "The Revolving Door," http://www.edmonds-institute.org/newdoor.html, accessed August 16, 2005. Information on International Policy Council from its website, http://www.agritrade.org/Outreach.htm, accessed August 17, 2005.

[16] Interview with author.

[17] Gurian-Sherman, Doug, "Holes in the Biotech Safety Net: FDA Policy Does Not Assure the Safety of Genetically Engineered Foods," report for Center for Science in the Public Interest(undated), available at cspinet.org/new/pdf/fda_report__final.pdf, accessed October 2006.

[18] "Plant Incorporated Protectants," U.S Environmental Protection Agency, available at http://www.epa.gov/pesticides/biopesticides/pips/index.htm, accessed October 2006.

[19] Sjoblad, Roy D., *et al.* "Toxicological Considerations for Protein Components of Biological Pesticide Products," *Regulatory Toxicology and Pharmacology* 15, 3-9 (1992)], cited in EPA Pesticide Fact Sheet, http://www.epa.gov/pesticides/biopesticides/ingredients/factsheets/factsheet_006481.pdf.

[20] Svitashev, S. K., W. P. Pawlowski, et al. (2002). Complex transgene locus structures implicate multiple mechanisms for plant transgene rearrangement," *The Plant journal: for cell and molecular biology* **32**(4): 433-445; Kohli, A., R. M. Twyman, et al. (2003). "Transgene integration, organization and interaction in plants." *Plant Molecular Biology* **52**(2): 247-258, from Quist, David, Chimeras in Context, doctoral dissertation, University of California, Berkeley, 2004.

[21] McCabe, M. S., U. B. Mohapatra, et al., "Integration, expression and inheritance of two linked T-DNA marker genes in transgenic lettuce," *Molecular Breeding* (1999), 5(4): 329-344; ,Srivastava, V., V. Vasil, et al. "Molecular characterization of the fate of transgenes in transformed wheat (Triticum aestivum L.)," *Theoretical & Applied Genetics* (1996) 92(8): 1031-1037; Spencer, T., J. O'Brien, et al. "Segregation of transgenes in maize," *Plant Molecular Biology* (1992) 18: 201-210; Cho, M., H. Choi, et al., "Inheritance of tissue-specific expression of barley hordein promoter-*uidA* fusions in transgenic barley plants," *Theoretical & Applied Genetics* (1999) **98**: 1253-1262, from Quist, ibid.

[22] Knudson, Tom, Edie Lau and Mike Lee, "Seeds of Doubt, Part Two: Globe-trotting genes: Welcome or not, modified strains pop up in crops

near and far," June 7, 2004, *The Sacramento Bee*, available at http://www.sacbee.com/static/live/news/projects/biotech/c2_1.html, accessed October 2006.

[23] Interview with author.

[24] Fritsch, Peter and Timothy Mapes, "In Indonesia, Tangle of Bribes Creates Trouble for Monsanto," *The Wall Street Journal*, April 5, 2005, available at http://online.wsj.com/PA2VJBNA4R/article_print/SB111264380462297385.html, accessed October 2006.

[25] Jaeger, Carlo, Ortwin Renn, Eugene Rosa, Thomas Webler, Risk, Uncertainty, and Rational Action (Earthscan 2000), p 168.

[26] Wiener, Jon. "Cancer, Chemicals and History," *The Nation*, Feb. 7, 2005, at http://www.thenation.com/doc/20050207/wiener, accessed October 2006.

[27] "Biotechnology at 25," University of California, Berkeley Library, available at http://bancroft.berkeley.edu/Exhibits/Biotech/25.html, accessed October 2006.

[28] Shorett, Peter, Paul Rabinow, Paul R. Billings, "The Changing Norms of the Life Sciences," *Nature Biotechnology*, Vol. 21, p 123-125 Feb. 2003.

[29] Lenzer, J. "Scandals have eroded US public's confidence in drug industry," British Medical Journal, BMJ 2004;329:247 (31 July), doi:10.1136/bmj.329.7460.247

[30] Beck, Ulrich, *Risk Society: Towards A New Modernity* (SAGE Publications Ltd., 1992), p54.

[31] Wagner, W.E. "The Science Charade in Regulatory Risk Assessments," presentation at the Society for Risk Analysis Annual Meeting 2004.

[32] See the 1978 Supreme Court ruling in *Vermont Yankee v. NRDC* and later cases of *Strycher Bay v. Karlen* (1980) and *Baltimore Gas & Electric v. NRDC* (1983); see also Rosebaum, 1974; Henderson and Pearson, 1978, Mandelker, 1981; Wenner, 1982., cited by Thomas Sander for the Saguaro Seminar, Kennedy School of Government, Harvard University.

[33] Bardach, Eugene, and Lucian Pugliaiesi. "The Environmental Impact Statement and the Real World," 1977, *Public Interest* 49 (Fall): 22-38, cited in Sander, ibid.

[34] Lester, James P., *Environmental Politics and Policy: Theories and Evidence* (Duke University Press, 1995), p. 214, cited in Sander, ibid.

[35] Sander, ibid.

[36] Daubert v. Merrell Dow Pharmaceuticals, Inc. 113 S.Ct. 2786 (1993).

CHAPTER 11. PUTTING PIGS TO THE TEST

[1] Stern, Paul C. and Harvey V. Fineberg, editors, *Understanding Risk: Informing Decisions in a Democratic Society*, Committee on Risk Characterization, National Academy Press, Washington, D.C. 1996, p 156

[2] Ibid.

[3] Fischhoff, B., "Scientific management of science?", *Policy Sciences* 33: 73-87, http://www.springerlink.com/content/g237j41611205l31/, Kluwer Academic Publishers, accessed October 2006.

[4] "Technical Guidance Document On the Use of Socio-Economic Analysis in Chemical Risk Management Decision Making," OECD Document ENV/JM/MONO(2002)10, OECD Environmental Directorate, Joint Meeting of the Chemicals Committee and the Working Party on Chemicals, Pesticides and biotechnology, March 15, 2002.

[5] "Managing Risks to the Public: Draft for consultation," Her Majesty's Treasury, October 2004, available at http://www.hm-treasury.gov.uk/consultations_and_legislation/greenbook_consultations/consult_greenbook_index.cfm, accessed October 2006.

[6] Douglas, Heather, "Inserting the Public Into Science," Sabine Maasen and Peter Weingart (eds.), *Democratization of Expertise? Exploring Novel Forms of Scientific Advice in Political Decision-Making* (Kluwer Academic Publishers, 2003).

[7] *Understanding Risk*, p 179

[8] "S Korea To Mass-Produce Pig Organs For Human Transplants," Dow Jones Newswires, June 1, 2004.

[9] Westphal, Sylvia Pagán, "Stolen transgenic pigs become sausages," July 25, 2001, *New Scientist*, available at http://www.newscientist.com/article.ns?id=dn1074, accessed October 2006.

[10] Jasanoff, Shiela, "Bridging the Two Cultures of Risk Analysis", *Risk Analysis* 1993, Volume 13, No. 2, p 123-129.

CHAPTER 12. WHAT THEN SHALL WE DO?

[1] Fischhoff, B., W Bruine de Bruin, U Guvenc, D Caruso, L Brilliant, "Analyzing disaster risks and plans: an avian flu example," *Journal of Risk and Uncertainty* (2006) 33: 133-151.

[2] Margolis, Robert M and David H Guston, "Origins, Accomplishments, and Demise of OTA," in *Science and Technology Advice for Congress*, Morgan and Piha, editors (RFF Press, 2005).

[3] A comprehensive list of publications of the Office of Technology Assessment are available at http://jya.com/otapub.htm, accessed October 2006.

[4] U.S. Congress, Office of Technology Assessment, *Difficult-to-Reuse Needles for the Prevention of HIV Infection Among Injecting Drug Users--Background Paper,* OTA-IW-H-103 (Washington, DC: U.S. Government Printing Office, October 1992).

[5] La Porte, Todd M., ibid.

[6] Goho, Alexandra, "Tiny Trouble: Nanoscale materials damage fish brains," April 3, 2004, *Science News*, Vol. 165, No. 14, p. 211, accessed October 2006, http://www.sciencenews.org/articles/20040403/fob1.asp.

[7] Ball, Philip, "Nanoparticles in sun creams can stress brain cells," *news@Nature.com*, June 16, 2006, available to subscribers at http://www.nature.com/news/2006/060612/full/060612-14.html, accessed October 2006.

[8] Weiss, R., in *Washington Post* (April 2006), as cited in *Nanotechnology: A Research Strategy for Addressing Risk*, July 2006, Wilson Center, available at http://www.nanotechproject.org/67/7-19-06-nanotechnology-a-research-strategy-for-addressing-risk, accessed October 2006.

[9] *Nanotechnology: A Research Strategy for Addressing Risk*, ibid.

[10] Conversation with author, October 2006.

[11] "Synthetic biology is ..." statement from http://syntheticbiology.org, accessed October 2006.

[12] "About Synthetic Genomics," available at http://syntheticgenomics.com/about.htm, accessed October 2006.

[13] Davies, Kevin, "Synthetic Biologists Assemble Codon Devices," June 2, 2005, *Bio IT World*, available at http://www.bio-itworld.com/newsitems/2005/05/06-02-05-news-codon-devices, accessed October 2006.

[14] "Synthetic Biology: SB2Declaration," available at http://syntheticbiology.org/SB2Declaration.html#Resolutions

[15] Caruso, Denise, "Risk: The Art and the Science of Choice," white paper for Rockefeller Foundation Global Inclusion Program, The Hybrid Vigor Institute, October 2002, p. 10-11, available from http://hybridvigor.net/publications.pl?s=health, accessed October 2006.

[16] Rejeski, David, "Making Policy in a Moore's Law World," *Ubiquity*, Volume 4, Issue 42, Dec. 17-23, 2003, Association of Computing Machinery, available at http://www.acm.org/ubiquity/interviews/v4i42_rejeski.html, accessed October 2006.

[17] "GMO-free regions and areas in Europe," available at GMO-free Europe, http://gmofree-europe.org, accessed October 2006.

[18] "Spotlight: GM Food Safety: Are Government Regulations Adequate?" Pew Initiative on Food and Biotechnology, 2004, available at http://pewagbiotech.org/buzz/display.php3?StoryID=42, accessed October 2006.

INDEX